Meteorites and Their Parent Planets

Meteorites and Their Parent Planets provides an engrossing overview of a highly interdisciplinary field – the study of extraterrestrial materials. The second edition of this successful book has been thoroughly revised, and describes the nature of meteorites, where they come from, and how they get to Earth. Meteorites offer important insights into processes in stars and in interstellar regions, the birth of our solar system, the formation and evolution of planets and smaller bodies, and the origin of life. Harry McSween's writing is accessible to scientists and nonscientists alike. He gives equal emphasis to the meteorites themselves and to what has been learned about the asteroids, comets, and planets from which they were derived.

The first edition was immensely popular with meteorite collectors, scientists and science students in many fields, and amateur astronomers. In this second edition all the illustrations have been updated and improved, many sections have been expanded and modified based on discoveries in the past decade, and a new final chapter on the importance of meteorites has been added. Everyone with an interest in meteorites will want a copy of this book.

Harry Y. McSween, Jr., is a professor and former head of the Department of Geological Sciences at the University of Tennessee. After graduating from The Citadel (B.S., Chemistry, 1967) and the University of Georgia (M.S., Geology, 1969), he served as an Air Force pilot during the Vietnam era. In 1977, he received a Ph.D. in Geology from Harvard University. Since that time, McSween has been a member of the faculty at the University of Tennessee and a NASA Principal Investigator, with research interests in meteorites and planetary geology. He has served on numerous NASA and U.S. National Academy of Sciences advisory panels that deal with research on extraterrestrial materials and space exploration and is a science team member for the *Mars Pathfinder, Mars Global Surveyor, Mars 2001 Orbiter*, and *Athena* spacecraft missions. He is also past president of the Meteoritical Society, an international organization devoted to the study of meteorites and planetary science. Besides the first edition of this book, McSween has previously published *Geochemistry: Pathways and Processes* (Prentice-Hall, 1989, coauthored with S. M. Richardson), *Stardust to Planets: A Geological Tour of the Solar System* (St. Martin's Press, 1993), and *Fanfare for Earth: The Origin of Our Planet and Life* (St. Martin's Press, 1997). He lives with his family in Knoxville.

Also by Harry Y. McSween, Jr.

Stardust to Planets: A Geological Tour of the Solar System

Fanfare for Earth: The Origin of Our Planet and Life

Geochemistry: Pathways and Processes

Meteorites and Their Parent Planets

Second Edition

Harry Y. McSween, Jr.

CAMBRIDGE
UNIVERSITY PRESS

PUBLISHED BY THE PRESS SYNDICATE OF THE UNIVERSITY OF CAMBRIDGE
The Pitt Building, Trumpington Street, Cambridge, United Kingdom

CAMBRIDGE UNIVERSITY PRESS
The Edinburgh Building, Cambridge CB2 2RU, UK   http://www.cup.cam.ac.uk
40 West 20th Street, New York, NY 10011-4211, USA   http://www.cup.org
10 Stamford Road, Oakleigh, Melbourne 3166, Australia
Ruiz de Alarcón 13, 28014 Madrid, Spain

First published 1999
Reprinted 2000

Printed in the United States of America

Typeset in Poster Bodoni 9.25/13 and Stone Serif in LaTeX 2_ε [TB]

A catalog record for this book is available from the British Library

Library of Congress Cataloging-in-Publication Data

McSween, Harry Y.
    Meteorites and their parent planets : thoroughly revised / Harry
Y. McSween. – 2nd ed.
       p.   cm.
    Includes index.
    ISBN 0-521-58303-9. – ISBN 0-521-58751-4 (pbk.)
    1. Meteorites.   I. Title.
QB755.M465   1999
523.5'1 – dc21   98-26494
                                                              CIP

ISBN 0 521 58303 9 hardback
ISBN 0 521 58751 4 paperback

Dedicated to the McSween clan

To my mother Frances, one of the most creative people I know
To my father Harry, epitomy of responsibility, who is sorely missed
To my sister Lucille, for strength, courage, and laughter
To my brother Dick, for a sense of perspective and fair play

Contents

Preface to the Second Edition

I acknowledge a fascination with jigsaw puzzles. Having spent countless hours assembling thousands of irregularly shaped pieces of cardboard into tabletop works of art, I have now graduated to three-dimensional puzzles. This interest is shared by many other people, of course, as evidenced by the rich variety of disassembled brainteasers on store shelves. But puzzles are sometimes more than games. For the scientists who study meteorites, puzzles are our shared profession.

Meteorites – the odd chunks of extraterrestrial rock and metal that periodically pelt our planet – are samples without geologic context. They arrive on Earth without labels identifying their places of origin, and piecing together enough information to make intelligent guesses about the identities and the properties of their parent bodies takes all the skill and patience of puzzle construction. But even in the best of cases, the assembled clues are fragmentary.

Solving the puzzling mysteries of meteorite parent bodies is of more than passing importance. Reconstructing this missing geologic context affords a way to examine the workings of worlds different from our own, bodies shaped by processes sometimes familiar and at other times bizarre, insights that offer an alternate vision of what might have been. Moreover, the identification of plausible or likely parent bodies for specific meteorite types reveals information on the organizational structure of our cosmic neighborhood and on the orbital gyrations of its residents. Meteorites also provide important details about the formative processes of the solar system itself, but these records have often been overprinted by parent body events. It would be folly to embark on a meteorite-based exploration program of the early solar system without first knowing how to recognize and compensate for the effects of parent body processing.

The first edition of this book was published in 1987. In preparing this thoroughly revised second edition, I have been astounded at how much has happened to meteoritics during the intervening time. The isolation and the characterization of stardust in chondritic meteorites, recognition of new classes of meteorites in Antarctic collections, the first spacecraft encounters with asteroids, new models that address the thermal and collisional histories of meteorite parent bodies, application of high-resolution chronometers to date early solar system events, a controversial hypothesis that a Martian meteorite contains evidence for extraterrestrial life, calculations revealing how meteorites trek from parent bodies to Earth, *Hubble Space Telescope* images of battered planetesimal – these are just a few of the exciting discoveries that have revolutionized this field. In addition to consideration of many new topics, virtually all the diagrams in this second edition have been revised and redrafted, and many new images are included.

The organization of the book follows that of the first edition, with pairs of chapters describing certain meteorites and their possible parent bodies. These paired sections are then followed by a chapter describing how meteorites are extracted from their parent bodies and placed in orbits that intersect that of the Earth. Finally, a new chapter discussing the importance of meteorites in unraveling the origin and evolution of the solar system has been added. Each chapter ends with an annotated bibliography for readers who wish to explore these subjects in more detail.

Meteoritics is a highly interdisciplinary field, requiring some knowledge of geology, chemistry, physics, and astronomy. Each of these disciplines, of course, has its own vocabulary. I have attempted to minimize the introduction of technical terms, but it is not possible to omit them entirely. Definitions of **boldfaced** words in the text are given in the Glossary at the end of the book. Readers who are unfamiliar with mineralogy may also wish to consult the Appendix of Minerals.

Any discussion of a subject this complicated may not always reach consensus; in such cases I have had to rely on my own prejudices. It is easy to read too much into meteorites, and in this book I may have been more guilty of that than most. I also take responsibility for simplifying the subject by omission of some aspects of meteoritics when these do not provide insights into different concepts or processes. I apologize in advance to my many friends and colleagues in meteoritics who may feel that their contributions have not been appropriately acknowledged. An unhappy circumstance of writing at this level is that the norms of scientific referencing cannot be followed. I appreciate the many individuals who have helped me find illustrations (acknowledged in figure captions). And finally, I thank my family for understanding this strange compulsion to write and not begrudging the many hours I spend doing it.

1 Introduction To Meteorites

Just before noon on November 7, 1492, a huge triangular stone hurtled out of the sky, crashing into a wheat field outside the walled city of Ensisheim in Alsace (now France). A curious crowd soon gathered around the meter-deep hole and, with great effort, hoisted the blackened rock out of the ground. The onlookers immediately began chipping off pieces of the object and carrying them away, thinking they were good-luck talismans. Hearing of this, the chief magistrate sought to protect the unusual stone by having it hauled into the city and placed at the door of the parish church. Several weeks later, the Emperor Maximilian happened to travel through Ensisheim and learned about the strange visitor. Thinking that the stone might be an omen of divine protection, he ordered that it be preserved in the church as testimony to this miraculous event. In accordance with medieval custom, it was fixed to the wall with iron crampons to prevent it from either wandering at night or departing in the same violent manner by which it had arrived. This 127-kg specimen, minus some fragments removed for museum collections during the ensuing five centuries, still resides in Ensisheim today (see Figure 1.1).

From Veneration to Disbelief

Meteorites were pelting the Earth, of course, long before there were people to observe them. More than a dozen fossilized meteorites have been found embedded in an Ordovician (approximately 475 million years old) limestone in Sweden. Even the Ensisheim incident was not the first account of the fall of a meteorite. Ancient chronicles from China and Greece document two independent meteoritic events occurring circa 650 B.C., and earlier, although less definite, records of meteorite falls from Crete extend as far back as 1478 B.C. Summerian texts from around 1900 B.C. describe a precious sample of metallic iron that is thought to have been meteoritic. The fall of an object from the sky naturally was (and still is) a dramatic occurrence, and it is understandable that such events were described in detail and that the recovered stones were venerated. The Suga Jinja Shinto

Figure 1.1: Meteorite falls were commonly given religious significance by ancient peoples. This woodcut carved in medieval times shows the fall of a meteorite near the town of Ensisheim (now in France) in 1492. The meteorite is illustrated as breaking through the clouds and also lying in a wheat field outside the town. On recovery, the 127-kg stone was placed in the local church. Like many important events of this time, the fall and its effect on the local populace were immortalized in verse, and this drawing decorated a broadsheet containing the poem. Except for one other meteorite in a Shinto shrine in Japan, the Ensisheim stone is the oldest preserved meteorite that was actually seen to fall.

shrine in Nogata, Japan, has kept a fist-sized meteorite as a treasure of the religious center for more than a thousand years. Its date of fall – May 19, 861 A.D. – is recorded in old literature as well as on the lid of the ancient wooden box in which it has been stored. A Russian meteorite fall in 1584 was apparently memorialized by the renaming of the town near which it fell (*Tashatkan*, literally stony arrow). The ancient Greeks and Romans enshrined and worshiped meteorites, which they called *betyls* (Hebrew for home of God). Coins were commonly struck to commemorate the fall of betyls, and many of these bear figures of the objects mounted in temples. According to Titus Livius, one meteorite (a conical object known as the needle of Cybele) was even conveyed in a royal procession from its impact site to Rome, where it was revered for another 500 years. These objects were also valued by the Egyptians and have been discovered entombed with the pharaohs in pyramids. Prehistoric American Indians transported meteorites long distances and sometimes buried them in crypts. One meteorite discovered in a burial ground of the Montezuma Indians in Casas Grande, Mexico, was wrapped like a mummy. For a number of years, the East African Wanika tribe worshiped a stone that fell in 1853, having anointed it with oil and clothed it. When enemies burned their village, the Wanikas concluded that their meteoritic deity was a poor protector, so they sold it to missionaries. In the early nineteenth century in Texas,

Comanche Indians placed a large meteorite at a point where several trails met, and Indian travelers passing by deposited offerings of beads and tobacco.

Even in more modern times, meteorites are sometimes accorded religious significance. The sacred black Kaaba Stone (its name means the right hand of God on Earth), to which Moslems in Mecca pay homage, is reported to have fallen from the sky, and many believe it to be meteoritic. Hindu religious literature states that meteorite falls herald important events, and in India it is reported that representatives of the Geological Survey must hurry to the site of any observed fall if they wish to collect the meteorite before it is enshrined by the local citizenry.

Early chronicles sometimes described objects other than stones falling from the sky. In 77 A.D., Pliny the Elder, in the same work that catalogued fallen stones enshrined in ancient temples, also listed fallen artifacts such as bricks, wool, and even milk. A 1513 book by Diebold Schilling contained a hideous painting of blood and flesh raining down on Rome during an apparition of Halley's comet, and descriptions of fallen stone axe heads and sharks' teeth were included in later publications. This uncritical assortment of fact and fancy about objects falling from the sky hindered acceptance of the reality of meteorites for centuries.

Most ancient philosphers viewed meteorites as heavenly bodies that had somehow been freed from their celestial moorings and had tumbled to Earth. This explanation was not universally accepted, however; Aristotle considered them to be atmospheric phenomena. In this regard he was prescient in expressing the view of scholars in later centuries, many of whom contrived to explain meteorites by atmospheric processes. Here is a typical example by W. Bingley, written in 1796:

It is but a trite observation to say, that the clouds make frequent visits to the waters of the earth, from which they usually carry away large quantities of that element, and with it, no doubt, the substances (even with some of the fish) which form the beds....It is self-evident, that the streams which ascend with the clouds are sometimes clear as crystal, at other times thick and muddy. When the latter is the case then it is that these substances may be concreted; and, by some extraordinary concussion in the atmosphere, return to the earth.

Others argued that meteorites were terrestrial rocks that had been struck by lightning, an explanation that spawned the popular name thunderstones for these objects. Yet another hypothesis held that meteorites were violently spewed from erupting volcanoes, either as lofted chunks of rock or as fine ash that coagulated in the atmosphere. This view was bolstered by the 1794 fall of a shower of stones near Siena, Italy, just eighteen hours after an eruption of Mt. Vesuvius. An account of the eruption and its possible relationship to the fallen stones was published by Sir William Hamilton, the English ambassador at Naples, who reported erroneously that the meteorites closely resembled the volcanic rocks of Vesuvius.

Many scientists, however, discounted altogether the idea that stones could fall from the sky. Their skepticism stemmed from the common conviction that no small objects existed in interplanetary space. As expressed by the preeminent physicist Isaac Newton:

To make way for the regular and lasting motions of the planets and comets, it's necessary to empty the heavens of all matter, except perhaps some very thin vapours, steams, or effluvia, arising from the atmospheres of the Earth, planets, and comets.

So strong was the belief in an empty cosmos that eyewitness accounts of meteorites plunging to Earth were ridiculed. After the fall of a meteorite was witnessed and described in a document notarized by the mayor and 300 citizens of Barbotan, France, in 1791, scientist Pierre Berthollet lamented

How sad it is that the entire municipality enters folk tales upon an official record, presenting them as something actually seen, while they cannot be explained by physics nor anything reasonable.

Meteorites could not begin to attract serious scientific scrutiny until such attitudes were dispelled.

The Early Days of Meteoritics

Meteoritics is the name given to the scientific study of meteorites. The father of this discipline was undoubtedly Ernst Chladni, a German physicist and lawyer (see Figure 1.2). Chladni's pioneering contribution was a small, 63-page book published in Riga, Latvia, and Leipzig, Germany, in 1794. In it he argued that meteorites, at least those composed mostly of metallic iron, certainly fell from the sky and were extraterrestrial in origin. Interestingly, he arrived at this astute insight before he had ever actually seen a meteorite. Chladni believed that these objects were once small bodies, traveling through space, that had come under the influence of the Earth's gravitational force. From the evidence for their intense heating and their compositional differences from terrestrial rocks, as well as the implausibility of other explanations, he proposed a relationship between such meteorites and atmospheric **fireballs** (words given in bold-faced type are defined in the Glossary at the end of the book). He correctly

Figure 1.2: Ernst Chladni, the father of meteoritics, as immortalized in a copper engraving.

surmised that air friction heated objects traveling at high speed through the atmosphere, producing an incandescent glow.

Chladni's idea that meteorites were extraterrestrial objects rather than atmospheric phenomena or volcanic ejecta amounted to scientific heresy, and his well-reasoned arguments were not immediately persuasive to his contemporaries. Resistance to this hypothesis lingered in part because of scientific conservatism, but also because most meteorites were stones rather than chunks of metal and thus were at least superficially similar to terrestrial rocks. Although at first Chladni's book received a chilly reception, the timing of its publication was perfect. Almost in immediate answer to its critics, a large stony meteorite fell in 1795 into the village of Wold Cottage, England. This event was important in refuting other currently popular mechanisms for the formation of meteorites (such as lightning or condensation in clouds), because the fall occurred out of a clear, blue sky. A specimen of the Wold Cottage meteorite eventually reached a young but highly respected British chemist, Edward Howard, who decided to perform a detailed analysis. Mineral chemistry was rather primitive in the eighteenth century, so that discriminating between meteorites and terrestrial rocks was challenging, but the field was progressing rapidly by the 1790s. Howard's careful study, done in collaboration with Jaques de Bournon, a French mineralogist exiled in England after the revolution, resulted in one of the first precise descriptions of a stony meteorite. In 1802, Howard reported concentrations of the element nickel in small grains of metal that de Bournon had separated from this stone. Nickel had earlier been analyzed in the metal of iron meteorites. Chladni's logical arguments had begun to persuade a number of scientists that iron meteorites were extraterrestrial, and this chemical link between irons and stones cleared the way for the interpretation that all meteorites had similar origins.

After these two pivotal publications, changes in the attitude of the scientific community began to occur fairly rapidly. In 1803, a number of eminent French scientists, convinced by reputable eyewitness accounts of meteorite falls and by their own confirmations of Howard and de Bournon's chemical findings, threw their prestige behind the proposition that meteorites were extraterrestrial objects. Remaining skepticism was silenced several months later when the town of L'Aigle, France, was peppered by a shower of no less than 3,000 stones. The French Minister of the Interior commissioned Jean-Baptise Biot, a physicist and one of the youngest members of the French Academy of Sciences, to investigate the L'Aigle incident. His 1803 report is commonly considered to be a turning point in the recognition of the authenticity of meteorites as extraterrestrial objects. In contrast to previous, rather dry scientific reports on meteorites, Biot's paper was dramatic and exciting. That notwithstanding, its impact was made possible by the careful research of his predecessors, and the importance accorded to this work in recent times may have been overestimated.

Within a decade of the appearance of Chladni's book, his hypothesis that meteorites were extraterrestrial had won general acceptance, and the science of meteoritics was launched. This is not to imply, however, that resistance to this new idea had totally vanished, especially outside of

Europe. For example, America's scientifically literate president, Thomas Jefferson, voiced skepticism about the authenticity of stones falling from the sky, even after hearing reputable reports of a meteorite fall in Weston, Connecticut.

From the beginning of the nineteenth century forward, meteoritics progressed steadily to an exacting and highly interdisciplinary field. The discovery of several asteroids within a few decades of the publication of Chladni's book demonstrated convincingly that interplanetary space was not empty. Marked advances in analytical chemistry and metallurgy occurred during the middle of the nineteenth century, and the invention of the petrographic microscope opened a new era of mineral identification. Developments during the twentieth century have revolutionized the field, and many of these are explored in the remainder of this book.

Properties of Meteorites

A **meteoroid** is a natural object of up to approximately 100 m in diameter that is orbiting in space. A **meteor** is the visual phenomenon associated with the passage of a meteoroid through the Earth's atmosphere. A **meteorite** is a recovered fragment of a meteoroid that has survived transit through the Earth's atmosphere. Meteorites are named for the geographic localities in which they fall or are found. As a consequence, meteoritics is endowed with a heritage of exotic place names that add to the allure of these objects. New meteorite names must be approved by the Nomenclature Committee of the **Meteoritical Society**, the international organization for research on meteorites.

I have already alluded to the existence of several types of meteorites – irons and stones. **Iron meteorites** consist almost entirely of nickel–iron metal alloys, whereas stony meteorites are composed mostly of silicate and oxide minerals, although many also contain small metal grains. In a third category are **stony–iron meteorites**, which have nearly equal proportions of metals and silicates. Early and widely used classification schemes, such as those developed by the European petrologists Gustav Rose in 1863, Gustav Tschermak in 1883, and Aristides Brezina in 1904, referred to irons as siderites, stony–irons as siderolites, and stones as aerolites. The stones can be further divided into two broad categories: chondrites and achondrites. A **chondrite** is a kind of cosmic sediment, an agglomeration of early solar system materials that has suffered little, if any, chemical change since its formation. In contrast, an **achondrite** is an igneous rock, the product of partial melting (accompanied by changes in chemical composition) and crystallization.

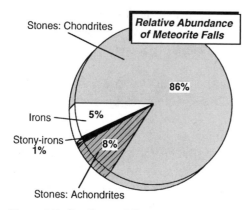

Figure 1.3: Significant differences in the proportions of meteorite types exist among fallen meteorites. The different areas of this pie diagram illustrate the relative abundances of chondrites, achondrites, irons, and stony–irons. The proportions of meteorite classes among finds are very different, being heavily weighted toward irons and stony–irons because their distinctive properties make them easily distinguishable from terrestrial rocks. The fall statistics may more accurately reflect the proportions of meteoroids orbiting in the vicinity of the Earth at the present time.

The proportions of iron and stony–iron meteorite **falls** (those seen to fall and then recovered) are very small, only approximately 5% and 1%, respectively, of the total number of meteorites collected. Of the remaining 94% stones, 86% are chondrites and 8% are achondrites (see Figure 1.3). These statistics may seem surprising to anyone who has looked at museum displays of meteorites, which are often dominated by irons and stony–irons. The ratio of irons to stones among meteorite **finds** (those recovered meteorites that were not observed to fall) is much larger than that among falls, because irons survive terrestrial weathering processes better and are more readily recognized as something unusual by nonscientists who stumble on them. Irons also tend to be larger and more spectacular in appearance than stones, so museum exhibits are often biased toward them. The ratio of meteorite types in falls probably accurately reflects the proportions of objects reaching the Earth from space during the past century, but it is doubtful that these statistics have any profound significance for the relative abundances of meteoroid types in space. It might be more informative to have data on the relative masses of meteorite types, but even mass ratios of different meteorite types must fluctuate over periods of millions of years.

Meteorites come in all sizes, but there is a marked tendency for irons to be larger than stones. The biggest meteorite thus far discovered is Hoba (Namibia), a block of nickel–iron metal weighing 55,000 kg (see Figure 1.4). Because it was proclaimed a national monument by the South

Figure 1.4: Hoba, the world's largest meteorite, weighs approximately 55,000 kg. It still rests at its impact site in Namibia, where it is a national monument. This picture was taken in 1928 soon after the meteorite was discovered. As seen in the photograph, the depression surrounding the massive iron is eroded crust formed by terrestrial weathering, so the original meteorite was even larger. Photograph courtesy of Brian Mason (Smithsonian Institution).

African government in order to save it from the smelter, it still lies embedded in the limestone in which it was first discovered in 1928. Surrounding this meteorite is a layer of rusty weathered material that formed by terrestrial alteration of the outer part of the object. If a correction is made for the amount of metal in this weathered halo, the original meteorite may have had a mass of more than 73,000 kg.

In 1897 the American Naval officer and Arctic explorer Robert Peary transported three massive pieces of another large iron meteorite from Cape York, Greenland, to New York City, where they can still be seen in the American Museum of Natural History. These, along with a fourth piece now in Denmark, have a collective weight of approximately 58,000 kg and together represent the second largest meteorite.

The largest meteorite found in the United States is the Willamette (Oregon) iron, weighing 12,700 kg. The meteorite was discovered in 1902 on property owned by the Oregon Iron and Steel Company, but the finder secretly moved the meteorite to his own property. This extraordinary feat, which he accomplished by mounting the meteorite on an ingenious trolleylike contraption drawn by horses, required three months of effort.

Unimpressed by this initiative, the company successfully sued to repossess the meteorite after the finder began exhibiting it; this case established the legal precedent (at least in the United States) that a meteorite belongs to the owner of the land on which it falls.

There is a substantial list of iron meteorites weighing more than 4,000 kg, and two stony–irons are this large. In contrast, stony meteorites are rarely greater than 500 kg in mass. The largest stony meteorite known, a chondrite weighing 1,750 kg, fell as part of a shower of fragments in the Jilin Province, China, in 1976 (see Figure 1.5). Some other stones have been found on the ground as groupings of related fragments that in a few cases may collectively weigh as much as a ton. Because they surely do not travel in space as closely grouped individual chunks, they must have been formed by the disruption of larger masses on impact or during atmospheric passage.

The smallest meteorites are collectively called **micrometeorites** or, more commonly, **interplanetary dust particles (IDPs)**. When the Earth passes through a region where dust is concentrated, such microscopic particles produce meteor showers as they burn up during atmospheric entry. Some dust particles actually reach the Earth's surface, but they are almost invariably melted. Miniature spherules collected in deep-sea sediments and in polar ice apparently formed in this way (see

Figure 1.5: The largest known stony meteorite, a chondrite weighing 1,750 kg, fell in the Jilin Province, China, in 1976. It was part of a large shower of stones produced by the breakup of a larger meteoroid during atmospheric transit. No one was injured.

Figure 1.6). A successful effort has been mounted to collect dust particles before they are destroyed or altered in the atmosphere by trapping them on sticky plastic plates located on the airfoils of high-altitude aircraft.

The Earth sweeps up perhaps 78 million kg of extraterrestrial material, most of which consists of micrometeorites, each year as it orbits about the Sun. The rate of fall for larger meteorites over the whole planet is more difficult to gauge with any degree of certainty. One calculation, extrapolated from eleven years of observations of meteors by a network of sixty automated cameras spread throughout western Canada, suggests that nine meteorites weighing at least 1 kg each fall annually over each million square kilometers of the Earth's surface and 54 kg of smaller meteorites fall within the same area during the year. This translates to one fall per square kilometer every hundred thousand years or so, on average. Another estimate, based on meteorites recovered within a desert region of New Mexico, is ten times greater. In any case, only a tiny fraction of this amount of material is actually recovered.

Meteorites that travel at high velocities through the atmosphere as single masses may develop distinctive shapes (see Figure 1.7). Because a meteor has no shields, such as those on spacecraft, to dissipate the heat generated by atmospheric friction, its leading edge will melt. Ablation of this molten rock or metal results in a smooth, featureless front surface. Some of the melt streams along the sides to the posterior surface, where it collects and solidifies into a roughly textured mass. The resulting conical shape is known as an **oriented meteorite**. If the shape of the projectile is not aerodynamically stable, it will rotate as it passes through the atmosphere. In this case its exterior will still melt, but no distinctive leading or trailing edges will be recognizable. A significant part (probably 30%–60% of the mass) of most meteoroids is lost to melting and ablation in the atmosphere. Calculations of rates of ablation suggest that the anterior surfaces of typical meteoroids lose 1–4 mm of material during each second of flight time. The surfaces of some meteorites are marked with depressions resembling thumbprints (called **regmaglypts**) that also form during atmospheric transit (see Figure 1.8). These are probably caused by the violent motion of air or selective melting and ablation of certain parts of the meteoroid. It is also common for incoming stones to break up into smaller pieces during atmospheric passage. Each of these will normally develop its own **fusion crust**, a layer of solidified melt **glass** coating the exterior. Such glassy surfaces are very thin, commonly less than a millimeter, except for solidified pools of melt on the trailing edges of oriented meteorites.

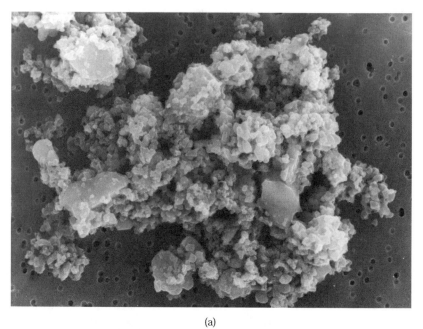

(a)

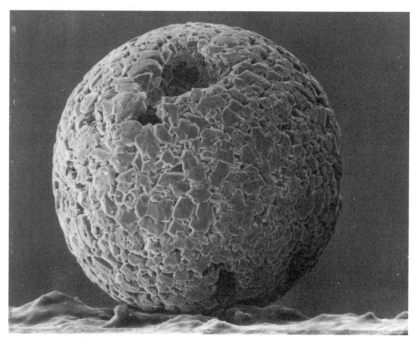

(b)

Figure 1.7: Partial melting of meteoroids due to frictional heating as they pass through the atmosphere may produce objects with unusual shapes. The cantelope-sized Bruno (Canada) iron meteorite shown here has six faces that were sculpted by ablation during atmospheric transit. The fine lines visible on the meteorite surface are frozen streams of melted fusion crust. Photograph courtesy of the Smithsonian Institution.

The external appearance of these objects produced by flight through the atmosphere often serves to identify them as meteorites. However, their internal compositions are also distinctive. Meteorites contain no elements that are not already present in terrestrial rocks, but these are combined in some cases to form unusual compounds. The minerals schreibersite (iron nickel phosphide), oldhamite (calcium sulfide), osbornite (titanium

←——

Figure 1.6: *(facing page)* (a) This tiny micrometeorite, viewed through an electron microscope, is only a few hundredths of a millimeter across. The fluffy aggregate of small crystals was collected in the upper atmosphere by a U-2 aircraft. (b) This spherule, recovered from deep-sea sediments, is also only a fraction of a millimeter in diameter. It is a solidified droplet, formed by melting of a micrometeorite (perhaps like the one above, although somewhat larger) during its passage through the atmosphere. Photomicrographs courtesty of Don Brownlee (University of Washington).

Figure 1.8: The heat built up during rapid deceleration and the violent air movements around a falling meteoroid may produce thumbprintlike depressions called regmaglypts. In this view, the Baszkówka (Poland) chondritic meteorite rests on its smooth trailing edge (during flight) and numerous regmaglypts can be seen on the upper, frontal surface. The meteorite is 30 cm in diameter. Photograph courtesy of Stanislaw Speczik (Museum of the Polish Geological Institute, Warsaw).

nitride), and sinoite (silicon nitrogen oxide), among others, have been recognized only in meteorites. However, these are relatively uncommon, and the major mineralogy of most meteorites is comfortingly familiar to geologists. Most of these minerals are silicates, such as olivine (magnesium iron silicate) and pyroxene (magnesium iron calcium silicate). Another common mineral that occurs in both meteorites and terrestrial rocks is plagioclase, a sodium–calcium aluminosilicate feldspar. The oxide minerals chromite (chromium iron oxide) and magnetite (iron oxide) are also common to both occurrences. Troilite (iron sulfide), cohenite (iron carbide), and several forms of nickel–iron metal (kamacite and taenite) are fairly abundant in meteorites but are extremely rare in terrestrial rocks and ores. Nearly 300 different minerals have been identified in meteorites, but the great bulk of most meteorites are composed of the minerals mentioned above. (An appendix of some of the most important meteoritic minerals is presented at the end of this book.)

A Fiery Passage

On the evening of November 8, 1982, the Donahue family of Wethersfield, Connecticut, settled down for a quiet evening of watching television when they were jolted by an apparent explosion, later described as "like a truck coming through the front door." In actuality, a stony meteorite had punctured the roof of their home. The family quickly discovered a large hole extending up through their ceiling into the roof. Fearing that there had been an explosion, they called the fire department. Within a few minutes, firefighters dispatched to the scene had found the intruder, a grapefruit-sized stone. The fire marshal later reported that the object was cool to the touch. The meteorite had plunged into the house at a 65° angle from the horizontal, finally coming to rest on the dining room floor. Six more smaller meteorite fragments were later found in the Donahue's vacuum, which Ms. Donahue had used to tidy up the mess before the media arrived.

One account of the Wethersfield fall was provided by an observer who was five miles away at the time:

It was close to 9:20 P.M., and I was jogging. I was headed straight west and saw a flash like lightning, and the entire sky was lit. I looked up, and about five degrees northwest of the zenith a very white and large object about the size of a basketball appeared. I thought it was space junk and saw fragments come off the eastern rim of it. The object stayed in one position and did not move – it was a head-on look. It could easily cast shadows. Six or more pieces fell from it in varying sizes. It then disappeared, and it never moved. I kept on jogging and began counting, anticipating a sound so I could judge the distance. I counted between 30 and 50 seconds and heard a series of six or more rifle shots going off in the direction of Wethersfield.

Although meteorites do not usually crash into houses, what made this incident unusual, at least to oddsmakers and the news media, was that another meteorite had fallen in the same small town just a few years earlier (also necessitating a roof repair). The 1982 fall occurred in the early evening, at approximately 9:20 P.M. local time. In this regard it was also somewhat unusual, because twice as many fireballs are reported in the hours after midnight. The bias in time occurs because more meteoroids are encountered in the direction in which the Earth moves in its orbit, and the morning side faces the direction of the planet's motion. Other than these unusual circumstances, the 1982 Wethersfield fall appears to have been rather typical. Let us examine some of the phenomena associated with this fall more critically.

The maximum velocity of any meteoroid orbiting in the solar system is 42 km/s, because objects with higher velocities would not have elliptical (closed) orbits and thus would escape the Sun's gravitational grasp. The Earth's orbital velocity is 30 km/s, so a head-on collision could potentially occur at up to 72 km/s. This extreme velocity is highly unlikely, however, because the meteoroid would have to travel in a **retrograde orbit**, moving in a direction contrary to practically all the bodies of the solar system. A meteoroid traveling in a **prograde orbit** should enter the atmosphere at substantially less than 42 km/s, as the Earth's velocity must be subtracted from that of a meteoroid overtaking the Earth from behind. This is consistent with the velocities measured for most fireballs, which range from 10 to 30 km/s. At altitudes below 100 km or so, the air density becomes great enough to create appreciable friction. This causes meteoroids to decelerate; it also accounts for their melting and ablation. Air resistance eventually slows most meteoroids to a **terminal velocity** of a few hundred meters per second, at which point the body has lost all its **cosmic velocity** (that which it had in space) and is free falling under the influence of the Earth's gravity. However, large meteoroids (those with weights exceeding approximately 10,000 kg) will not be affected as much by atmospheric resistance and can continue all the way to the Earth's surface while still retaining at least part of their cosmic velocities (see Figure 1.9). Besides its mass, a meteoroid's atmospheric deceleration will also be affected by its entry angle and drag coefficient (controlled by size and shape).

The jogger who witnessed the Wethersfield fireball reported that it was intensely white and that its glow was extinguished before it struck the ground. Some observed fireballs have been bright enough to illuminate 100,000 km^2 of the Earth's surface (approximately the area of Alaska). The decrease in emitted light from the Wethersfield meteor was a consequence of its rapid deceleration and the subsequent reduction in air friction; at terminal velocity, meteors are no longer luminous. When a meteoroid penetrates the atmosphere at greater than supersonic speed, it creates a shock wave as the air in front of the object is compressed. This effect is similar to that produced by a bullet or a supersonic jet aircraft. Such shock waves may generate sound phenomena resembling thunder or detonations, as noted at Wethersfield. Many observers near the points of meteorite impacts have reported consistent patterns of sounds: whistling or buzzing like that produced by falling bombs, tearing or rumbling noises, and detonations analogous to those produced by supersonic aircraft.

The 1982 Wethersfield meteor was seen to break up into six or more pieces during its flight. Nearly half of all fireballs are observed to fragment,

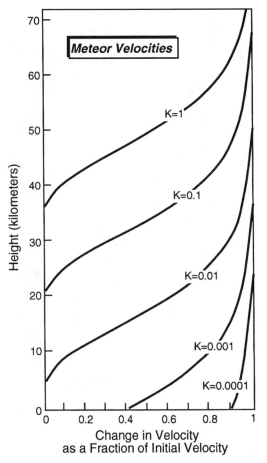

Figure 1.9: Massive meteoroids penetrate much deeper into the atmosphere than smaller ones before being slowed by air friction. The curves in this diagram illustrate the changes in velocity for meteoroids of various masses, shapes, and entry angles between the meteoroid's trajectory and the Earth's surface. These variables are incorporated into the parameter K, which is a numerical value directly proportional to the drag coefficient (related to the shape of the object) and inversely proportional to the mass and the sine of the angle of incidence. The horizontal scale gives the change in velocity at any height as a fraction of the initial entry velocity. Large meteoroids with low K values will penetrate the atmosphere without much loss of velocity, whereas small ones with K values of 0.01 or greater will lose all their cosmic velocity at some point and finally free fall to Earth under gravity.

most at altitudes between 12 and 30 km. The fall of a disrupted meteoroid may result in a **strewn field** of many individual meteorites. In the normal case of an oblique approach, the pattern on the ground delimited by the recovered meteorites will be elliptical (see Figure 1.10). The largest masses tend to carry greater distances than their smaller relatives, because

Figure 1.10: Many meteoroids break up during atmospheric passage and fall to Earth as showers of numerous fragments. In the normal case of an oblique approach, the impacting meteorites will form a strewn field of elliptical shape. This aerial photomosaic shows the strewn field for a chondritic meteorite shower that fell in Saint-Robert (Canada) in 1994; the rectangles are grain fields and pastures. Twenty specimens of various sizes were recovered; their relative masses are illustrated by location dots of different sizes. The original object before breakup may have weighed 10,000 kg. The shower approached from the southwest, and the larger meteorites at the northeastern end of the strewn field decelerated more slowly and thus carried farther. Photograph courtesy of Alan Hildebrand (Geological Survey of Canada).

larger objects retain some of their cosmic velocity longer and thus travel farther. This idealized shape can be modified, of course, by repeated fragmentations in the air or other complicating factors.

The damage to the Donahue's home caused by the impact of the small Wethersfield meteorite was certainly not minor, but it was less than one might expect from a rock falling to the Earth from space. Meteorites striking the ground may excavate small cavities, but typically they penetrate only to depths nearly equal to their diameters. In fact, many meteorites are found practically sitting on the surface of unconsolidated soil or snow. This is, of course, due to the appreciable deceleration produced by atmospheric friction. A falling meteorite is still dangerous, but there are no credibly reported instances of anyone being killed in such an incident (see Figure 1.11). There is a record of a young girl who was struck by a falling meteorite in Juashiki (Japan) in 1927, but she was not

Figure 1.11: Although small meteoroids are decelerated by atmospheric friction, they still may be traveling at considerable speed on impact with the Earth. The meteorite in the right hand of the lady in the polka-dotted dress penetrated the roof of her garage in Benld, Illinois, in 1938. It continued through the roof, back seat, and wooden floorboards of the automobile in the background, bounced off the muffler, and ultimately lodged in the seat cushion. Photograph courtesy of Edward Olsen (Field Museum of Natural History).

seriously hurt. In 1954, another woman was badly bruised when the falling Sylacauga (Alabama) meteorite crashed into her home, and the Nakhla (Egypt) achondrite that fell in 1911 reportedly hit and killed a dog.

It may seem surprising that the Wethersfield meteorite was cool to the touch, because a significant portion of it must have been heated to melting temperature during its fall. Most of the molten material, however, had already been lost to ablation before impact, leaving only a very thin rind of quenched fusion crust. Although in some cases fusion crusts may still be warm, the interiors of these objects certainly are not. Meteorites have been stored in the deepfreeze of space for eons, and atmospheric heating does not signficantly affect their interiors because heat conduction in stones or even irons takes much longer than the minute or so required for atmospheric transit. The Colby (Wisconsin) and Dharmasala (India) meteorites are reported to have been quickly coated with frost, even though both fell on hot days in midsummer.

Desert Meteorites, Frozen and Otherwise

Antarctica is the coldest, windiest, highest above sea level, most arid, and most inaccessible continent in the world. It is at the same time a place of awesome natural beauty and glacial nastiness. On its rocky basement rests approximately 90% of the Earth's ice, and within this is a treasure trove of meteorites. This unexpected source of extraterrestrial material was uncovered by accident in 1969, when Japanese glacial geologists stumbled on nine meteorites exposed on bare ice near the Yamato Mountains. Thinking that the meteorites were pieces of the same fall, the scientists collected and sent them home for study. To everyone's amazement, the nine specimens included examples of four different classes of meteorites. Japanese geologists returned to the Yamato ice fields in 1973 specifically to search for meteorites, and in 1976 American field parties dedicated to this purpose began exploration of the interior flank of the Transantarctic Mountains. Other expeditions have been mounted by Germany and by a consortium of European countries. These groups have now collected thousands of meteorite specimens from a number of regions in the Antarctic (see Figure 1.12). The most productive areas have been the Yamato ice fields in Queen Maud Land and the Allan Hills in Victoria Land.

In these locations vast swarms of meteorites are scattered over stretches of bare ice. The mystery of the manner by which meteorites are concentrated in these areas, called **meteorite stranding surfaces**, has now been at least partly unraveled by William Cassidy of the University of

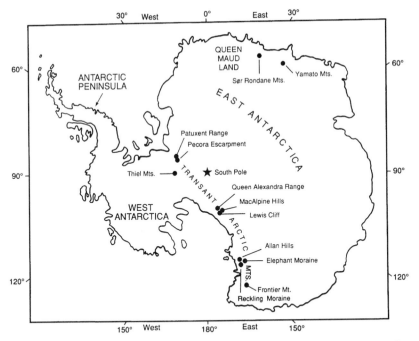

Figure 1.12: Meteorite concentrations have been found in Antarctica at a number of sites in the Transantarctic Mountains by U.S. scientific teams and in the Rondane and Yamato Mountains in Queen Maud Land by Japanese expeditions. Meteorites recovered from these sites in the past few decades number in the thousands. Figure by William Cassidy (University of Pittsburgh).

Pittsburgh, leader of the U.S. Antarctic meteorite expeditions for many years. The polar ice sheet, which covers 12 million km², provides an ideal catchment area for meteorites falling over long periods of time. These objects are quickly frozen into the thickening ice. This ice sheet reaches a thickness of approximately 4,000 m near the center of the Antarctic continent, and in places its weight has depressed the underlying rocks to below sea level. The tremendous mass of the ice causes it to squeeze downward and flow outward toward the edges of the continent. The ice and its enclosed meteorites creep along at rates of several meters per year, moving toward an ultimate fate of breaking off into the sea as icebergs. At some locations, however, the horizontal motion of the ice is arrested by obstructions. Effective barriers to impede the ice movement are provided by mountains (often not recognizable as such because they are nearly covered or even overrun by the ice cap; the tips of mountains poking through the ice are known by the Eskimo term *nunataks*). Nearly stagnant ice is uplifted by forces pushing against the obstruction and eroded away by

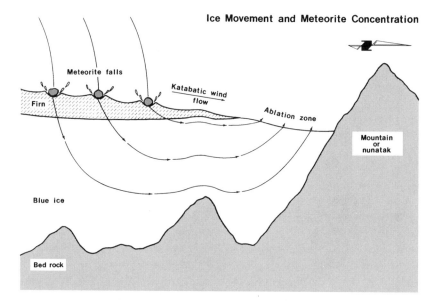

Figure 1.13: This diagram schematically illustrates the meteorite concentration mechanism in the Allan Hills and possibly other locations in Antarctica. All over the continent, meteorites fall into firn and are ultimately frozen into the accumulating blue ice. The ice sheet flows outward toward the edges of the continent unless it meets an obstruction such as a nunatak. Stagnant ice behind such a barrier will undergo ablation by katabatic winds, and meteorites will accumulate as successive layers of ice are exposed and removed in this way. Diagram courtesy of NASA.

evaporation or ablation due to strong winds, producing stranding surfaces. The katabatic (descending) winds roar down the gently sloping ice sheet, clearing the stagnant areas of snow and producing a storm of dancing ice crystals that sandblast the ice surface. Measured ablation rates in such areas indicate that approximately 5 cm of ice are removed each year, continually exposing new ice at the surface. As each successive layer of ice, carrying an occasional meteorite, is uncovered and eroded away, the meteorite remains behind to join others and form a concentration (see Figure 1.13).

Moving glacial ice is normally blanketed by a layer of snow, called firn, that appears white. In contrast, stagnant ice that is being actively ablated is referred to as blue ice because of its distinctive coloration. Areas of exposed blue ice can be spotted in photographs taken by satellites orbiting in space, and this is where the search for Antarctic meteorites begins. Scientific teams are airlifted to blue ice fields already identified in satellite images, and detailed searches are carried out on the surface by snowmobile. The collection efficiency at some localities has been remarkable,

Figure 1.14: The dark color of meteorites provides a stark contrast to the Antarctic ice and snow and makes collection relatively easy. This chondritic meteorite was found by the author near Reckling Peak (Antarctica) in 1981. The stone is covered with dark fusion crust, but the lighter interior can be seen in the lower right-hand corner where the meteorite has been chipped. The counter shows the number assigned to this sample in the field. Photograph courtesy of NASA.

because the dark fusion crusts of even thumbnail-sized meteorites stand out in stark contrast to the bright icy background (see Figure 1.14).

The recovered meteorites are numbered, photographed, and collected in sterile plastic bags to prevent contamination of the samples. The Antarctic environment is clinically clean, and despite their residence in the enveloping ice, many specimens are less contaminated than most of those in museum collections. The still-frozen meteorites are shipped back to the NASA Johnson Space Center in Houston (by American field parties) or to the National Institute of Polar Research in Tokyo (by Japanese field parties). From these facilities, processed samples are sent to research laboratories all over the world for scientific study (see Figure 1.15). Eventually the American meteorites become part of the collection of the U.S. National Museum (Smithsonian Institution).

It is difficult to know how many of the individual meteorite specimens collected so far in Antarctica represent individual falls. By 1996

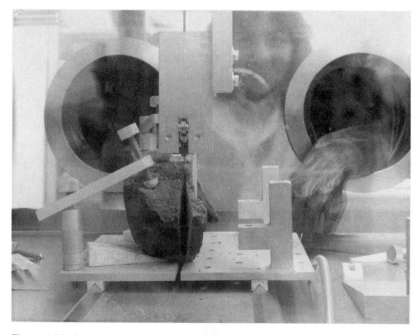

Figure 1.15: Meteorites recovered by U.S. expeditions to Antarctica are transported in the frozen state to the NASA Johnson Space Center in Houston, Texas. The meteorite processing laboratory provides curation and processing facilities similar to those accorded lunar samples returned by the Apollo missions. Here a technician cuts a slab through a large achondrite. The saw is enclosed within a glovebox that preserves a contamination-free environment. Photograph courtesy of NASA.

the number of specimens had grown to more than 15,000. Clearly, many of these must be fragments of the same meteorites (**paired meteorites** in the jargon of meteoritics), but the moving ice sheets can disperse them over a wide area. If only a fifth of the samples recovered so far represent individual falls, then at least 3,000 new meteorites have been added to our collections in less than three decades. This compares with approximately 2,600 catalogued meteorites in museums all over the world that have been collected over the past two centuries.

The Antarctic meteorites provide a difficult problem in nomenclature. Because of the high concentrations of meteorites within small areas and the scarcity of local geographic landmarks, the naming system used for other meteorites is not applicable, and a special system has been devised. An Antarctic meteorite is identified by some nearby landmark such as the Allan Hills (normally abbreviated in capital letters, as in ALH for this location), plus a number such as 84001 (the first two digits correspond to the year in which the expedition that found the sample arrived in

Figure 1.16: Desert meteorites are often easily noticed. This chondritic meteorite, weighing 1.75 kg, was discovered by a European field party in 1993 on the Jiddat al Harasis Plain of Oman, Saudi Arabia. Even though the meteorite's contrast with the surrounding limestone pebbles is considerable, nine sets of tire tracks within 20 m of the find testify to the low probability of meteorite discoveries by individuals not specifically looking for them. Photograph courtesy of EUROMET.

Antarctica, 1984 in this example, and the last three digits identify the specific sample in that year's collection).

Antarctic meteorites are especially interesting because they may have had long residence times on Earth. By using sophisticated radioactive clocks in these specimens (the mechanics of which are discussed in Chapter 8), we can estimate the elapsed times since they fell. Ages measured in this way for these meteorites range from a few thousand to more than 900,000 years; compared with Antarctic meteorites, even the Ensisheim fall occurred only yesterday. These recovered objects thus provide some perspective on the kinds and the relative abundances of meteorites that were falling to Earth before recorded human history. The Antarctic source has also afforded the opportunity of discovering new types of meteorites. A number of rare or even unique kinds of meteorites have been recovered, further adding to the excitement of opening this meteoritic treasure chest.

The annual precipitation rate in Antarctica is so low that it is classified as a desert. Other desert areas of the world, especially the Nullarbor Plain of Australia and parts of New Mexico in the United States, have traditionally been fruitful locations for meteorite finds. The low humidity allows meteorites to be preserved for long periods of time (a few have been on Earth for as much as 40,000 years), and minimal camouflaging by vegetation and high prevailing winds that scour the surface of small particles aid discovery. The Sahara is now emerging as a particularly productive area. For five years, beginning in 1989, 471 meteorites were collected in Algeria and Libya. The meteorites were recovered from so-called Regs, which are flat, almost featureless desert areas. Very few of these samples appear to be paired, although many are heavily weathered and rusty in appearance (see Figure 1.16).

Target Earth

We have already seen that massive meteoroids entering the Earth's atmosphere are not fully decelerated and retain some of their cosmic velocities all the way to the ground. Impacts of such large objects are fortunately rather rare events because such bodies are less abundant than smaller objects. When such **hypervelocity impacts** do occur, they produce **craters**. Massive meteoroids, or even larger orbiting bodies, called **asteroids**, carry tremendous amounts of kinetic energy and have the potential to cause great devastation. For example, a stony asteroid with a diameter of 2 km and traveling at 50 km/s would hit with an energy equivalent to 4 trillion tons of TNT. This nontrivial explosion would produce a crater 33 km in diameter and nearly 3 km deep. This example is admittedly an extremely large body, but it is a reasonable size for some of the asteroids that periodically are observed to approach the Earth's orbital path.

Impact craters produced by even larger and more energetic objects are found on the Earth's surface. Approximately 130 terrestrial craters are currently recognized, ranging in size up to several hundred kilometers in diameter and in age from a few thousand years to approximately 2 billion years. Meteoritic debris has been found at approximately a dozen such craters. We have already seen that iron meteorites tend to be larger than stones, and it is probably no coincidence that these craters are associated with irons and stony–irons.

Geologists have been slow to recognize impact craters and, until the past few decades, have been reticent to accept hypervelocity impact as the mechanism by which they formed. The first link between a sizable terrestrial crater and meteorite impact was forged by Daniel Barringer near the beginning of the twentieth century. Barringer owned the Arizona property containing Meteor Crater (then known as Coon Mountain), and his interest in the crater began as a mining venture. Impressed by the abundant iron meteorites found near the structure, he proposed to exploit the supposed large metallic mass buried beneath the crater floor. Over the years Barringer drilled a number of shafts, but never struck meteorite. It is now evident that in this and other large hypervelocity impacts, the projectiles are mostly demolished by fragmentation and melting. Fine metallic globules that appear to be remnants of the original impacting mass have been recovered from soil as far away as 15 km from Meteor Crater. In 1906 Barringer published his observations supporting the idea that a large projectile had excavated Meteor Crater. However, his hypothesis was mostly ignored, and not until the time of his death in 1929 did a sizable part of the geologic community accept his contention that meteorite impacts could produce craters on the Earth.

In partial defense of geologists, I should note that impact craters are not always as obvious as one might think. Many craters are disguised by having been filled with sediments and lakes. An impact origin for the New Quebec crater in Canada was not suspected until a prospector in 1950 noted, from an aerial photograph, the unusual circular shape of its contained lake. The 3.7-km-diameter basin was not substantiated as a crater until scientists examined the area and the rocks carefully. The New Quebec crater is only 10,000 to 20,000 years old and is a relatively fresh feature. If it went unrecognized for so long, one can imagine that identification of older, more deeply eroded craters might be difficult indeed. The eroded remnants of ancient craters are called **astroblemes**, literally star wounds. Where weathering and erosion are not pervasive, as on the Moon, the surface is pockmarked with craters of all sizes accumulated over billions of years. On such bodies without atmospheres, even small incoming meteoroids can produce hypervelocity impacts. At the other extreme is Venus, whose dense atmosphere screens out any asteroids smaller than a kilometer or so in diameter.

There are numerous criteria other than associated meteorites that can be used to ascertain whether or not a circular depression on Earth may be an impact crater. The nature of the rocks in the walls and on the floor of the crater provides important clues. The cavity is commonly filled with shattered rock, and localized puddles of melted rock may be buried deep in the crater bottom. In a fresh crater, the surrounding rocks bulge upward to form a raised rim, and in places original rock layers near the rim may even be folded back on themselves. The intense shock pressures in rocks outside the crater may form conical structures called shatter cones, which point toward the center of the impact. Such high pressures also produce diagnostic kinds of microscopic deformation in the target rocks, as well as transform certain minerals into more tightly packed crystalline forms or cause them to melt.

Meteor Crater (see Figure 1.17) remains the best-studied example of a hypervelocity impact crater. This bowl is 1.1 km in diameter and 150 m deep. However, it is now half filled with shattered and locally melted rock, and its original depth was greater. Rain and an ancient lake have also washed other unconsolidated sediments into the basin. The rock formations into which the meteorite impacted bulged upward to form the crater rim, and an overturned flap of sedimentary layers occurred on the eastern margin. The age of the crater is approximately 50,000 years, and the impacting meteorite is thought to have been roughly 30 m in diameter and weighed perhaps 100,000 tons. Some of the sandstones in the target area contain coesite and stishovite, very dense minerals formed from original quartz grains that were subjected to high shock pressures. The surrounding desert is littered with fragments of the Canyon Diablo iron meteorite, which in aggregate weigh more than 18,000 kg. Meteorite specimens found on the crater rim contain small diamonds, whereas those collected farther from the crater do not. The diamonds, also very dense minerals, were produced from graphite by shock in fragments that may have been spalled off the rear of the impacting projectile. Meteorite specimens from the surrounding plains presumably were smaller pieces broken off the meteoroid in the atmosphere, and thus these may have escaped the extreme shock conditions of hypervelocity impact.

The cratering process can be visualized in three stages: compression, excavation, and modification. During compression, the kinetic energy of the meteorite is transferred to the target rocks in the form of compressional energy. During the excavation stage, the highly compressed rocks subsequently relax, causing crushed material to be propelled out of the crater. This material forms a blanket of deposited **ejecta** around the crater. Very large craters are subsequently modified in shape as rocks surrounding the crater fault or slump inward to produce central peaks or concentric rings (see Figure 1.18). All these stages may occur within seconds or minutes of the impact.

Hypervelocity impacts could have been responsible for some of the great extinctions of living species that punctuate the geologic record. The best known of these extinctions marked the end of the Mesozoic era some 65 million years ago and could have resulted in the demise of the dinosaurs as well as many other organisms. In all, approximately a fourth of the known families of animals disappeared within a short time interval. In 1978, Luis and Walter Alvarez (father and son) and several of their colleagues at the University of California, Berkeley, made a serendipitous discovery: A thin layer of brown clay, sandwiched between Mesozoic and Cenozoic limestones at Gubbio, Italy, contained a remarkably high

Figure 1.17: Meteoroids that are so large that they are not slowed appreciably during passage through the atmosphere hit with tremendous force. These hypervelocity impacts excavate craters, of which Meteor Crater (also called Barringer Crater), Arizona, is a classic example. The crater, shown here in an oblique aerial photograph and a cross-section sketch, is a simple bowl with a diameter (D) of approximately 1.1 km. Its apparent depth (d_a) is several hundred meters, approximately half of its true depth (d_t) because it is partly filled with shattered rock. Holes were drilled by the mining company that owned this property in hopes of locating a large mass of buried iron meteorite, but they were unsuccessful. Photograph courtesy of the Smithsonian Institution; schematic cross section by Richard Grieve (Geological Survey of Canada).

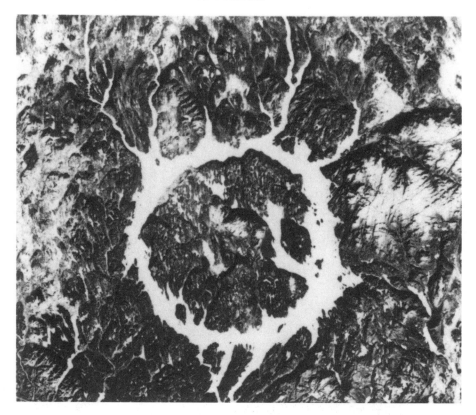

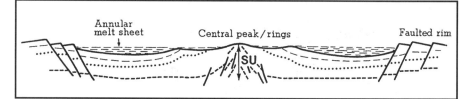

Figure 1.18: Very large impacts produce complex craters with central peaks formed by rebound of the compressed materials. This satellite photograph shows the Manicouagan Crater (Canada), one of the largest impact craters in North America. The annular lake is 70 km in diameter and surrounds an inner plateau containing a structural uplift (SU) in the center. Photograph courtesy of LANDSAT; schematic cross section by Richard Grieve (Geological Survey of Canada).

concentration of the element iridium. Similar spikes in the abundance of iridium were found in sediments deposited at the same time in other locations around the world. This element is greatly enriched in meteorites relative to terrestrial rocks, and the Alvarez team suggested that its high abundance might represent the chemical signature of a massive meteorite impact. This chemical fingerprint apparently occurs worldwide in both continental and marine sediments of the same age. It seems likely that a large impact lofted enough pulverized rock and dust into the atmosphere to block out sunlight, thereby lowering global temperatures and greatly restricting plant photosynthesis. Disruption of the food chain or alteration of climate could have snuffed out many land and sea creatures over a short period of time. Other mechanisms for climate alteration have also been attributed to such a death-dealing impact. The Alvarez hypothesis fomented a great deal of interest, and sometimes consternation, among scientists and the public alike. A large crater of just the right age, however, was nowhere to be found. Then, just a few years ago, a huge crater was discovered in Mexico, off the tip of the Yucatan Peninsula. The 300-km-diameter bowl, named Chicxulub (translated from the Mayan language as tail of the devil) after a nearby village, lies buried under a kilometer of sediments and is half submerged beneath the waters of the Gulf of Mexico. Geophysical surveys and drilling have confirmed its impact origin and allowed scientists to estimate that the impact melted as much as 20,000 km^3 of rock in just a few seconds. Droplets of this melt were splashed outward for thousands of kilometers, to be incorporated into boundary clays in Haiti and the western United States. No meteorites, however, have yet been found that would allow recognition of the identity of this interplanetary intruder.

Too Close for Comfort

The Earth resides within a swarm of asteroids and comets, collectively called **near-Earth objects (NEOs)**. The orbits for a few hundred NEOs having estimated diameters of 1 km or larger are known (see Figure 1.19), but the entire NEO population may contain perhaps 3,000 such objects. The likelihood that one of these objects will strike the Earth in the near future is remote but nonzero, and its effects could be disastrous. Indeed, during our lifetime, the chance that an impact large enough to destroy food crops on a global scale and possibly threaten civilization is roughly 1 in 10,000. The recognition that Earth-approaching objects pose a finite hazard to life has led to a flood of new discoveries and NASA funding of systematic surveys for potentially hazardous objects. Their orbital motions bring most NEOs relatively near the Earth every few years, and it is during such approaches that they are normally discovered.

Asteroids in near-Earth space are categorized as **Amor, Apollo,** or **Aten** objects, depending on whether their orbits lie outside that of the Earth,

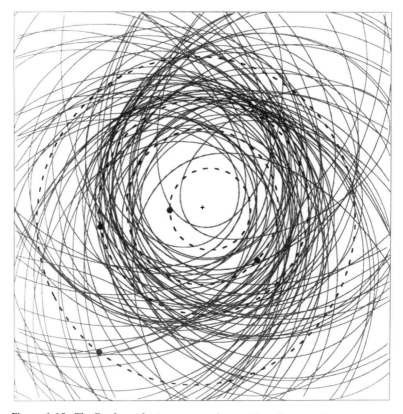

Figure 1.19: The Earth resides in a swarm of asteroids and comets that may constitute impact hazards. The orbits for 100 NEOs with diameters of 1 km or larger represent only a fraction of the estimated population in this size range. To construct this figure, all orbits have been rotated into the Earth–Sun (ecliptic) plane. The Sun is the small cross in the center of the field; also shown as dashed curves are the orbits of the terrestrial planets (Mercury through Mars, with increasing distance from the Sun), with their positions on January 1, 1997, illustrated as dots. Figure constructed by Richard Binzel (Massachusetts Institute of Technology).

overlap that of the Earth with periods of greater than one year, or overlap that of the Earth with periods of less than one year, respectively. Some NEOs are undoubtedly spent comets, having been altered from their original states by numerous close passes to the Sun.

The earliest-discovered NEOs were given the names of the more active and sometimes erotic figures of Greek mythology, such as Apollo, Amor, Eros, and Adonis. The number of known NEOs has now far outstripped the available list of mythical names, and new asteroids, whether near-Earth or not, are named after whomever and whatever the discoverer wishes (subject to approval by the International Astronomical Union).

The name is normally preceded by a number corresponding to the order of discovery, as in 887 Alinda or 1915 Quetzalcoatl. Just after discovery, asteroids get a preliminary designation of a year followed by two letters (except I and Z), the first indicating the half-month of observation (A indicates the first half of January and B the last) and the second the order of discovery within the half-month. The temporary name, as in 1993 FA, is eventually replaced with a permanent moniker when the object is found again and its orbit is confirmed. In contrast to asteroids, comets are named after the discover.

Radar images, obtained at the Arecibo Observatory in Puerto Rico and the NASA Jet Propulsion Laboratory's Goldstone Antenna in California, have provided informative views of a few NEOs. For example, asteroid 4179 Toutatis is a dumbbell-shaped body whose unusual rotation causes it to tumble end over end (see Figure 1.20). A few NEOs are the most readily accessible extraterrestrial bodies for exploration by spacecraft, requiring less energy than is necessary for reaching the surface of the Moon. In 1999, the *Near-Earth Asteroid Rendezvous* (*NEAR*) spacecraft is scheduled to encounter and orbit 433 Eros, one of the largest NEOs, measuring 14 km × 14 km × 40 km.

Many meteorites are undoubtedly derived from NEOs. However, NEOs are efficiently removed from the solar system by collisions or gravitational interactions with the planets on time scales of 10–100 million years, only a tiny fraction of the age of meteorites. We thus infer that today's temporary NEO population must be continually resupplied from other sources. These sources are the **parent bodies** of meteorites.

Meteorite Parent Bodies

Early conjectures about meteorite parent bodies were as imaginative as those about the origins of the meteorites themselves. In his ground-breaking book, Ernst Chladni recognized that meteorites did not form in space as the small chunks that periodically fall to Earth, and he postulated that they must have been derived from the breakup of larger bodies "due to an external impact or an internal explosion." His ideas were already controversial enough that he wisely refrained from speculating about which heavenly bodies these were.

Soon after the publication of Chladni's book, German astronomer Heinrich Olbers suggested that meteorites might be ejecta from volcanoes on the Moon. At first, Chladni joked about the idea, noting that a shower of beans had allegedly fallen in Spain, and that twelve bushels were collected, cooked, and found to be delicious, so a lunar eruption might have

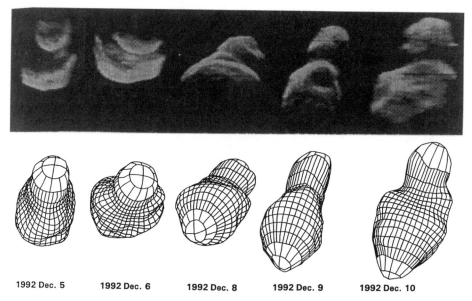

1992 Dec. 5   1992 Dec. 6   1992 Dec. 8   1992 Dec. 9   1992 Dec. 10

Figure 1.20: The near-Earth asteroid 4179 Toutatis, as imaged by ground-based radar, is a potato-shaped object with an unusual, tumbling rotation. The 6-km-long asteroid was at first thought to be two co-orbiting objects. The radar images (top) were made in 1992 by Steven Ostro and his colleagues at the NASA Jet Propulsion Laboratory when Toutatis was only 2.2 million miles from Earth. The drawings (below), by Philip Stooke of the University of Western Ontario, correspond to the radar image orientations, with dates of image acquisition shown. Photograph courtesy of NASA.

catapulted a grocery store into space. A few years later, however, he had second thoughts and, for a time, endorsed the idea of lunar origin. This idea has been resurrected repeatedly during the ensuing two centuries, mostly notably in recent times by Harold Urey, an American Nobel Laureate. The idea was abandoned again when lunar samples returned during the Apollo program turned out to be distinct from achondrites known at that time. Urey has finally been vindicated, however, by the discovery of a handful of lunar meteorites in Antarctica.

Chladni's original suggestion – that meteorites might be fragments of an exploded planet – also has had a long history. Well before Chladni's ideas were published, Titius von Wittenburg in 1766 had already noted a regularity in the spacings between planets. This sequence is a geometric progression, commonly expressed as

$$r = 0.4 + 0.3(2^n),$$

where r is the radius of the orbit of any planet whose numerical order outward from the Sun is n. Astronomers considered this regularity in planetary locations to be an important, primary feature of the solar system, especially after Sir William Herschel discovered the planet Uranus in 1781 at an orbital distance that closely fitted this equation. Using this rule, Johann Bode forcefully argued that a planet was missing between Mars and Jupiter, and the relationship came to be known as the Titius–Bode law.

In 1801 the Sicilian astronomer Giuseppi Piazzi accidentally discovered a small planetary object located approximately at the Titius–Bode distance for the missing planet. Excitement coursed through the scientific establishment as Bode and others asserted that Piazzi's observation filled this planetary gap. In reality, Piazzi had discovered the first asteroid, 1 Ceres. However, the next year, Heinrich Olbers discovered yet another body, the asteroid 2 Pallas, in the same orbital vicinity. Olbers was, by this time, a convert to Chladni's hypothesis that meteorites were fragments of an exploded planet, and he logically concluded that Ceres and Pallas were also fragmental remnants of this body. This idea became very popular and remained so for a long time. In 1847, French geologist Adolphe Boisse published a conceptual cross-section of the meteorite parent planet, with meteorites arranged concentrically in the order of their densities. Even in the twentieth century, there has been no shortage of advocates for the position that all meteorites could have been derived from the disruption of one large, stratified planet. In 1943, Harvard geologist Reginald Daly provided a detailed reconstruction of the lost planet, which by that time had been christened Phaeton, after the mythological character who rode a fiery chariot across the sky. Daly surmised that iron meteorites formed the core of this 6,000-km-diameter planet, succeeded outward by a mantle of stony–irons and a crust of stony meteorites.

The next year, a theory for the origin of the solar system formulated by the respected Russian academician O. J. Schmidt was suggested to be inconsistent with the exploded planet model. Instead, Schmidt proposed that the asteroids represented an arrested stage of planet formation and had never been assembled into a large body. This idea neatly circumvented the thorny problem of how to disintegrate a planet.

Several other variations on the small parent body theme also gained adherents at one time or another. In 1910 the Italian astronomer Giovanni Schiaparelli, most remembered as the discoverer of the Martian canals, determined a close orbital relationship among meteor showers and some comets, and he proposed that meteors (and hence meteorites) were cometary debris. That idea has now been resurrected with the hypothesis that some IDPs come from comets.

Two Russian scientists, I. S. Astropovich and Vladimir Vernadsky, even considered the possibility that meteorites came from outside the solar system, having wandered in from the surrounding galaxy. In a subsequent paper, Vernadsky dismissed the asteroidal model as "an assumption based on seventeenth century ideas, alien to celestial mechanics and universal contemporary views." If the history of meteoritics teaches us anything, it is to be cautious in making definitive statements about meteorite origins.

The hypothesis that meteorites might be derived from a planet such as Mars, Venus, or Mercury is a recent suggestion. The difficulties encountered in extracting a rock from a planetary gravitational field are formidable, but apparently not prohibitive. So far, only meteorites thought to be from Mars have actually been recognized.

Identification of the parent bodies of meteorites remains today a contentious subject, but new observational techniques and spacecraft missions have shed further light on the problem. Moreover, the meteorites themselves carry information that is useful in reconstructing the histories of their parent bodies. To say too much at this point would give the story line away, but evidence bearing on the identities of meteorite parent bodies is examined thoroughly in Chapters 3, 5, and 7.

No One Knows Quite Enough

After the people of Ensisheim had secured the newly fallen meteorite in their church, they placed the following inscription near it:

Many know much about this stone, everyone knows something, but no one knows quite enough.

Despite the fact that we have learned a tremendous amount about meteorites in the ensuing five centuries, the veracity of this inscription remains unchanged. That, of course, is part of what makes meteoritics perpetually interesting. The remainder of this book is devoted to explaining some of what we do know (or currently surmise) about meteorites and their parent bodies. In the following chapters, we consider each of the main classes of meteorites separately, focusing on what has been learned about their properties, origin, and subsequent evolutionary history. For each meteorite group we also attempt to reconstruct and, if possible, identify its parent body or bodies. We then consider how these interplanetary wanderers got into the Earth's neighborhood. Finally, we explore some of the more profound secrets that meteorites carry.

Suggested Readings

Most of the references below are very readable without a sacrifice of quality. Some are more technical but still highly informative, and should be digestible by nonspecialists.

GENERAL

Dodd R. T. (1986). *Thunderstones and Shooting Stars: The Meaning of Meteorites*, Harvard U. Press, Cambridge, MA.

This nontechnical account provides excellent descriptions of meteorite types.

Heide F. and Wlotzka F., translated by Clarke R. S. Jr. (1995). *Meteorites: Messengers from Space*, Springer-Verlag, Berlin.

The fourth edition of a popular German book about meteorites for nonscientists.

Hutchison R. (1983). *The Search for Our Beginning: An Enquiry Based on Meteorite Research, Into the Origin of Our Planet and Life*, Oxford U. Press, London.

A lucidly written, nontechnical introduction to meteoritics.

Kerridge J. F. and Matthews M. S., eds. (1988). *Meteorites and the Early Solar System*, University of Arizona, Tucson, AZ.

A highly technical treatise containing fifty contributions from sixty-nine authors, all authorities in the field. Possibly the best available resource on meteorites, but not for the beginner.

Norton O. R. (1994). *Rocks from Space: Meteorites and Meteorite Hunters*, Mountain Press, Missoula, MT.

A richly illustrated book for beginners, with an especially useful section on recognizing meteorites.

Wasson J. T. (1985). *Meteorites: Their Record of Early Solar-System History*, Freeman, San Francisco, CA.

A more technical survey of meteorites, focusing on interpretations of their origins.

HISTORY OF METEORITICS

Burke J. G. (1986). *Cosmic Debris: Meteorites in History*, University of California, Berkeley, CA.

The only modern, book-length history of meteoritics, this exhaustively referenced work provides an authoritative historical reference.

Marvin U. B. (1992). The meteorite of Ensisheim: 1429–1992. Meteoritics **27**, 28–72.

A nicely illustrated, well-written account of the earliest witnessed meteorite fall in the western world for which the object is preserved.

Sears D. W. (1978). *The Nature and Origin of Meteorites*, Oxford U. Press, London.

A concise technical book on meteorites with an especially good section on the historical development of the science of meteoritics.

METEORITE CLASSIFICATION

Graham A. L., Bevan A. W. R., and Hutchison R. (1985). *Catalogue of Meteorites*, 4th ed., British Museum (Natural History), London, and University of Arizona, Tucson, AZ.

This bible of classified meteorites is an exhaustive compendium of information about the World's meteorite collections, mostly exclusive of Antarctic meteorites.

Grossman J. N. (1984). The Meteoritical Bulletin, No. 76, 1994 January: The U.S. Antarctic collection. Meteoritics **29**, 100–143; Grossman J. N. and Score R. (1996). The Meteoritical Bulletin, No. 79, 1996 July: recently classified specimens in the United States Antarctic Meteorite Collection (1994–1996). Meteoritics Planet. Sci. **31**, A161–174.

The best available compilations of classified Antarctic meteorites; no comparable listing is available for Antarctic meteorites in the Japanese collection.

METEORITE FALL AND RECOVERY

Brown P., Hildebrand A. R., Green D. W. E., Page D., Jacobs C., Revelle D., Tagliaferri E., Wacker J., and Wetmiller B. (1996). The fall of the St-Robert meteorite. Meteoritics Planet. Sci. **31**, 502–517.

A particularly complete description of the phenomena associated with a meteorite fall.

Halliday I., Blackwell A. T., and Griffin A. A. (1989). The flux of meteorites on the Earth's surface. Meteoritics **24**, 173–178.

A statistical treatment of fireball events detected by the Canadian camera network.

Zolensky M. E., Wells G. L., and Rendell H. M. (1990). The accumulation rate of meteorite falls at the Earth's surface: the view from Roosevelt County, New Mexico. Meteoritics **25**, 11–17.

A contrary estimate of meteorite flux based on meteorites recovered in one area.

ASTEROID DISCOVERY

Pilcher F. (1989). The circumstances of minor planet discovery. In *Asteroids II*, edited by R. P. Binzel, T. Gehrels, and M. S. Matthews, University of Arizona, Tucson, AZ, pp. 1002–1033.

This tabulation gives the discovery date and place and the discover's name for thousands of asteroids.

IMPACT CRATERS

Chapman C. R. and Morrison D. (1989). *Cosmic Catastrophies*, Plenum, New York.

An excellent review of the emerging scientific debate on the role of impacts in geologic history.

Grieve R. A. F. (1991). Terrestrial impact: the record in the rocks. Meteoritics **26**, 175–194.

This excellent summary includes a tabulation of the locations, sizes, and ages of known terrestrial craters.

Hodge P. W. (1994). *Meteorite Craters and Impact Structures of the Earth*, Cambridge U. Press, Cambridge.

The topic of impact cratering is not adequately described in this book, but it does provide a compilation with illustrations of terrestrial impact structures.

Melosh H. J. (1989). *Impact Cratering: A Geologic Process*, Oxford U. Press, London.

An authoritative, physics-based text on impact phenomena and crater formation, well illustrated with minimal mathematics.

DESERT METEORITES (HOT AND COLD)

Bischoff A. and Geiger T. (1995). Meteorites from the Sahara: find locations, shock classification, degree of weathering and pairing. Meteoritics **30**, 113–122.

A technical paper describing meteorites found in the desert.

Cassidy W., Harvey R., Schutt J., Delisle G., and Yanai K. (1992). The meteorite collection sites of Antarctica. Meteoritics **27**, 490–525.

Everything you want to know about finding meteorites in Antarctica.

2 Chondrites

Imagine a witness at the birth of the solar system, painstakingly observing and recording each event as it unfolds. What would such a recording be worth to science now? The origin and the early evolution of the Sun and planets are still, to a degree, shrouded in mystery, because there are no surviving witnesses or records – none, that is, but chondrites. The name of this important meteorite group derives from the ancient Greek word *chondros*, meaning grain or seed, a reference to the appearance produced by numerous small, rounded inclusions called **chondrules**. Chondrules are visible in the accompanying photograph of a cut slab of an Antarctic chondrite (see Figure 2.1). In this chapter we attempt to lift the shroud a bit and peek into the dark recesses of earliest solar system history. We do this by examining the record imprinted in chondrites. One might think that such chunks of rock would be mute witnesses, but nothing could be further from the truth.

Once upon a Time

If chondrites contain records of early solar system processes, they must be very old. But how old are they, and how does their age compare with that of the solar system itself?

The Earth is our most accessible source of solar system material, and its formation age should be approximately the same as that for the whole system. This, of course, presumes that there was no significant gap in time between the formation of the Sun and the planets, and we have no theoretical reason or evidence to suggest a significant hiatus. The only suitable clock for this determination uses naturally occurring **isotopes**. Atoms of a given element are distinguished from those of other elements on the basis of the number of protons (positively charged particles with mass) in their nuclei, but different atoms of the same element may contain different numbers of neutrons (nuclear particles that have mass but no charge). Atoms that differ in mass, that is, in the number of neutrons they contain, are called isotopes. Unstable (**radioactive**) isotopes

Figure 2.1: Chondrites take their name from the small stony spherules, called chondrules, that they often contain in abundance. The origin of these small crystallized droplets remains one of the most perplexing aspects of chondritic meteorites. This photograph shows the sawed face of a chondrite found in Antarctica in 1977. This object contains an interesting assortment of rounded chondrules of various sizes. The small block is a scale measuring 1 cm on a side. Photograph courtesy of NASA.

transform, by loss of protons, neutrons, and electrons, over time into more stable isotopes of other elements. The parent isotopes decay at fixed (and measurable) rates, allowing the age of any sample to be determined from analysis of the amount of the parent isotope that has decayed or the new (**radiogenic**) isotope that has formed. Unfortunately these radioactive clocks can be reset by any geologic events that cause heating. This situation is analogous to a tape recorder that automatically erases the existing tape each time a new program is recorded. Most radioactive clocks in terrestrial rocks are still suitable tools for geologic work, but they have been reset so many times that the formation age of the Earth cannot be determined from them directly. In a few places, such as parts of Australia and Greenland, very old rocks are found. The most ancient rocks recognized thus far, the Acasta gneisses of Canada, are almost 4 billion (U.S. billion, 10^9, here and throughout) years old. These are very old, to be sure, but not old enough. The radioactive dates that these rocks record are metamorphic (recrystallization of solid rock during heating) events, so the Earth must be older still. Actually, determining the age of the Earth is not as intractable a problem as it might seem, but in order to resolve this we must first determine the age of chondrites.

To illustrate the principle of a radioactive clock applicable to meteorites, we consider the decay of an unstable isotope (**radionuclide**) of the element rubidium, ^{87}Rb. The superscript 87 is the mass number of the isotope, equal to the sum of its protons and neutrons. Half of the radioactive ^{87}Rb in any sample will decay to ^{87}Sr, an isotope of strontium, in approximately 49 billion years. This time interval is the **half-life** of ^{87}Rb. Knowledge of this rate of decay, coupled with measurement of the amount of ^{87}Rb that has transformed into the new strontium isotope, will permit calculation of the age of this sample. The problem is a little more complicated, however, because not all the ^{87}Sr now in the meteorite was produced by the decay of ^{87}Rb – some was already there when the meteorite formed. The mathematical expression for this is

$$^{87}Sr_{now} = {}^{87}Sr_{original} + \left({}^{87}Rb_{original} - {}^{87}Rb_{now} \right),$$

where the terms in parentheses, corresponding to the amount of ^{87}Rb that has decayed, of course equal the amount of new ^{87}Sr produced.

The law governing radioactive decay states that the terms in parentheses, the original and the final amounts of unstable isotope, are related by $(e^{\lambda t})$. The term e is the number used as the base for natural logarithms (approximately 2.718), and in this expression it is raised to a power equal to the product of the rate of decay λ and the elapsed time t (which is what we want to know). This decay law can be written as

$$^{87}Rb_{original} = {}^{87}Rb_{now}(e^{\lambda t}).$$

Substituting this expression for ^{87}Rb$_{original}$ into the previous equation gives

$$^{87}Sr_{now} = {}^{87}Sr_{original} + {}^{87}Rb_{now}(e^{\lambda t} - 1).$$

^{87}Sr is not the only isotope of strontium present in the meteorite. ^{86}Sr also occurs, but it is not a decay product (that is, not radiogenic), so its proportion does not change with time. However, because the amount of ^{87}Sr increases as ^{87}Rb decays, the ratio ^{87}Sr/^{86}Sr increases as time passes. Dividing both sides of the previous equation by a constant, ^{86}Sr, will not affect the equality:

$$\frac{^{87}Sr_{now}}{^{86}Sr} = \frac{^{87}Sr_{original}}{^{86}Sr} + \frac{^{87}Rb_{now}}{^{86}Sr}(e^{\lambda t} - 1)$$

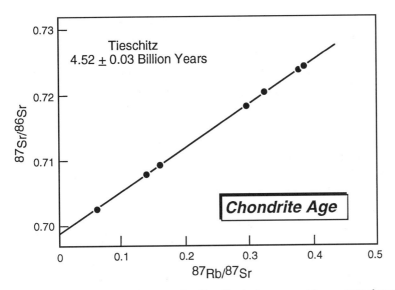

Figure 2.2: Precise measurement of radioactive isotopes provides a way to determine the age of a chondrite. In the method illustrated here, a radioactive isotope of rubidium (^{87}Rb) decays to a stable isotope of strontium (^{87}Sr), which mixes with already existing strontium isotopes (both ^{87}Sr and ^{86}Sr). The individual data points in this diagram represent minerals separated from the Tieschitz (Czech Republic) chondrite. The age of the meteorite (4.52 billion years) is calculated from the slope of the line defined by these points, which steepens as the elapsed time increases. Data obtained by Jerry Wasserburg and co-workers (California Institute of Technology).

Note that we can measure two of the foregoing ratios in the meteorite, $^{87}Sr_{now}/^{86}Sr$ and $^{87}Rb_{now}/^{86}Sr$. Typically, several separated fractions of the meteorite are analyzed, each fraction containing different minerals or at least different proportions of the same minerals. If we construct a graph plotting the two analyzed ratios for these mineral fractions versus each other, they will define a straight line, like that in Figure 2.2. Because all samples on this straight line have the same age, the line is called an **isochron**. The equation above is the mathematical expression of this line, in the standard form $y = b + mx$, where m (the slope of the line) is ($e^{\lambda t} - 1$) and b (the y intercept) is $^{87}Sr_{original}/^{86}Sr$, the initial strontium isotopic ratio.

Therefore, from the slope of the line in this diagram, we can determine the age of the meteorite, and we have circumvented the thorny problem of not being able to measure the amount of ^{87}Sr that was present in the original sample. As time passes, the slope of the isochron will become steeper. Isotopic data for mineral fractions of the Tieschitz (Czech Republic) chondrite analyzed by a **mass spectrometer** are shown in Figure 2.2. An age of 4.52 ± 0.3 billion years was calculated from the slope of this isochron. Similar ages have been determined from other chondrites, with the most precise ages clustering around 4.56 billion years. Approximately at this time, rubidium and strontium were locked into mineral grains as they crystallized, and the radioactive clocks in chondrites started ticking.

But what is the relationship of this age to that of the solar system? What evidence is there to argue against the notion that chondrites may have formed much later, say a billion years after the formation of the Sun and planets? To answer this question, we must turn to other isotopic systems. Unlike ^{87}Rb, some radioactive isotopes decay very rapidly. One example is an isotope of iodine, ^{129}I, which transforms to an isotope of xenon, ^{129}Xe, with a half-life of a mere 16 million years. Any ^{129}I in the early solar system would persist at detectable levels for only 100 million years or so, equivalent to approximately six half-lives. Thus, if chondrites incorporated live ^{129}I, they must have formed close to the time this isotope was created, which is thought to date the solar system's formation. Luckily the atomic abundance of iodine is far greater than that of xenon, so enough excess ^{129}Xe might be formed by radioactive decay to alter perceptibly the ratio of this isotope to other xenon isotopes. Morever, xenon is an element whose isotopic abundances can be measured with phenomenal precision. Figure 2.3 illustrates the relative proportions of xenon isotopes in the Richardton (North Dakota) chondrite. For comparison, the shaded areas show the isotopic composition of xenon in the Earth's atmosphere, probably a reasonable approximation to the average xenon isotope ratios of the solar system. For this experiment, the xenon gas is driven out of the meteorite by heating, and the excess ^{129}Xe is released along with iodine, implying that it was originally trapped as live ^{129}I. The former presence of this now-extinct radionuclide implies that chondrites formed at the same time as the rest of the solar system.

Not all short-lived radionuclides produce radiogenic isotopes that can be directly measured. However, in some cases other kinds of fossil evidence of now-extinct isotopes can be found in chondrites. For example, an isotope of plutonium, ^{244}Pu, with a half-life of only 82 million years, decays into an array of lighter isotopes. This process releases a great deal of energy, as is graphically illustrated by its unfortunate use in nuclear weapons. For plutonium trapped in minerals, this energy is imparted to the newly created particles, propelling them away through adjacent crystals at high velocities. These miniature projectiles leave tiny trails of destruction, called **fission tracks**, that we can enlarge to microscopically visible tracks by etching the crystals with acid. Some other long-lived radionuclides, especially those of uranium, produce similar tracks, because

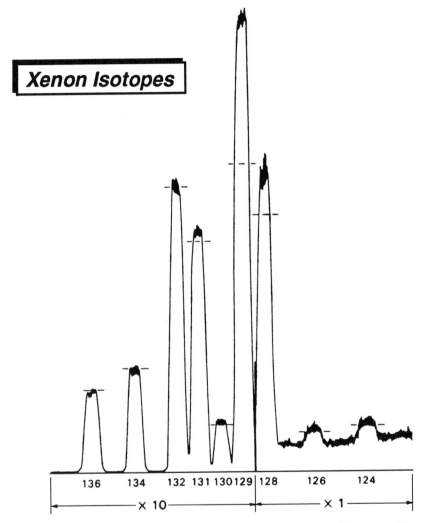

Figure 2.3: The idea that chondrites are as old as the solar system itself is suggested by the fact that they contain the decay products of isotopes that existed only fleetingly after the solar system formed. Peak heights in this diagram represent the amounts of the various isotopes of xenon (^{136}Xe, ^{134}Xe, etc.) relative to ^{132}Xe, in the Richardton (North Dakota) chondrite. The left-hand part of the spectrum has been enlarged ten times. The dashed horizontal lines are the relative amounts of different xenon isotopes in the Earth's atmosphere. The large excess in ^{129}Xe results from the rapid decay of a now-extinct isotope of iodine (^{129}I), which was present for only a few million years after the solar system's birth. Measurements by John Reynolds (University of California at Berkeley).

they eject radiogenic particles with comparable energies. However, the densities of tracks in chondrites are too great to be accounted for by decay of uranium alone. The extra fission tracks point to the prior existence of ^{244}Pu when the meteorite crystals formed.

To summarize, chondrites formed approximately 4.56 billion years ago, as determined from slowly decaying isotopic systems. The former existence of short-lived and now-extinct radionuclides and their incorporation into chondrites indicate that these meteorites must have formed within the first few million years of solar system history, commonly called the **formation interval**. Therefore we accept an age of approximately 4.56 billion years for the birth of the solar system.

This discussion of time began with a short discourse on why terrestrial rocks cannot be used to date the age of the Earth and, by inference, the solar system. Let us reconsider this question by examining uranium and lead isotopes. I have saved this discussion for last because the uranium–lead system is somewhat more complicated than the isotopic systems we have already considered. There are two radioactive isotopes of uranium (^{235}U and ^{238}U) that decay ultimately into different isotopes of lead (^{207}Pb and ^{206}Pb, respectively). By themselves, neither of these parent–child pairs in terrestrial rocks can be used to date the Earth's formation. However, this age can be determined if both systems are used simultaneously. It is possible to model the evolution of both lead isotopes through time if we know the initial isotopic proportions of lead and uranium. The initial isotopic compositions can be approximated if we assume that the Earth originally had the same isotopic ratios of uranium and lead that chondrites had. In 1956 geochemist Claire Patterson, of the California Institute of Technology, by using lead isotopes, first calculated a correct age of the Earth as 4.55 ± 0.7 billion years, essentially the same as that determined for chondrites.

Cosmic Chemistry and Chondrite Classification

The chemical composition of the solar system is often referred to as the **cosmic abundance** of the elements. The solar system's composition is really equivalent to the chemistry of the Sun, which contains most of the mass (more than 99%) of the whole system. The term cosmic abundance is therefore misleading, as it does not indicate the composition of the cosmos, but only a small part of it in which we are egocentrically interested. In fact, many other stars are known to have chemical compositions that are different from that of the Sun, and there is no way to estimate a truly representative cosmic composition.

The visible white-hot surface of the Sun is called the photosphere. The chemistry of the photosphere has been measured by astronomers from the absorption of certain wavelengths of energy by elements in their excited states. The abundances of most elements in the Sun have been determined to no better than ±40% of the amounts present, but this level of precision is sufficient for comparison with chondrites.

Chemical studies of chondrites have a long and rich history. The famous chemist Antoine Lavoisier was a member of a commission of the French Academy of Sciences who performed the first crude chemical analysis of a chondrite, published in 1772. Modern research on chondrite chemistry has progressed to a level of high precision as ever more sophisticated techniques have been brought to bear on the problem. As a result, chondrites are the most thoroughly and accurately analyzed natural materials, and the list of precisely analyzed elements encompasses the entire Periodic Table.

Figure 2.4 shows a plot of the abundance of each element in a chondrite versus that element's abundance in the solar photosphere. These data are plotted on a logarithmic scale, with the numbers referring to exponents to the base 10. The abundances of different elements vary over many orders of magnitude, and a logarithmic scale allows all these element concentrations to be plotted on the same diagram. The Sun obviously contains many more atoms of each element than does a small meteorite, but ratios of elements permit us to compare their compositions. In this figure, all measurements are referred to an arbitrary standard value, one million atoms of the element silicon. Exactly the same element ratios in the Sun and chondrites would produce points that fall along the diagonal line in Figure 2.4. The correspondence between the Sun and chondrites is very good, one might even say exceptional. No kind of terrestrial rock would show such an agreement because the chemistry of rocks changes each time they undergo geologic processing. The major discrepancies in Figure 2.4 involve the lightest elements (hydrogen, helium, carbon, nitrogen, and oxygen), which are consistently more abundant in the Sun. Under most conditions, these so-called **volatile elements** exist primarily as gases. Thus chondrites can be considered a sort of solar sludge, with compositions equivalent to the nonvolatile portion of the Sun. Because all elements can be analyzed much more precisely in meteorites than in the Sun, chondrite analyses are used to specify cosmic abundances of all but the most volatile elements.

Several other elements besides the volatiles also deviate slightly from the diagonal line. Lithium and, to a lesser extent, boron deviate the other way, that is, they are more abundant in chondrites than in the Sun. These

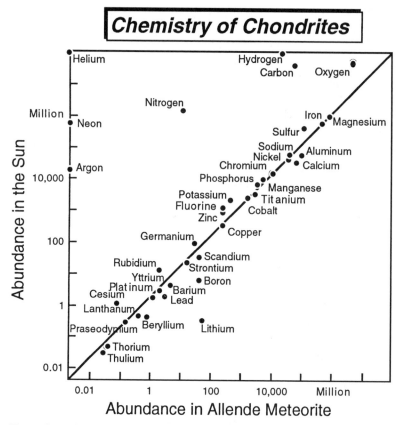

Figure 2.4: The chemical compositions of chondrites closely match that of the Sun, suggesting that such meteorites represent primitive materials that have survived without significant chemical change since the formation of the solar system. This diagram compares elemental abundances (atoms per one million atoms of silicon) in the Allende (Mexico) chondrite with those in the Sun. Both scales are logarithmic. A perfect correspondence is indicated by the diagonal line, and only volatile elements deviate significantly from the line.

are real differences outside the limits of analytical error. The discrepancies can be explained by the fact that lithium and boron are utilized in fusion reactions that power the Sun. Their solar abundances have been reduced during the past 4.56 billion years, so in this way chondrites actually record the chemistry of the ancient Sun (hence the primeval solar system) even better than does the present-day Sun.

Several different chemical groups of chondrites with distinct compositions have now been recognized. This may seem like a contradiction and
requires some elaboration. How can chondrites have cosmic compositions
and yet be different from each other? Remember that Figure 2.4 is a logarithmic plot spanning many orders of magnitude, so minor concentration
differences of a few percent would not even be noticeable. Chondrites are
classified into **groups**, which have a limited range of composition and
probably formed on the same parent body, and **clans**, which are sets of
groups that have properties in common and are inferred to have formed in
the same region of the solar system. The compositional clans of chondrites
we distinguish are the **ordinary chondrites** (so named because they are
the most abundant clan), the **carbonaceous chondrites** (always carbon bearing, but actually misnamed when it was believed that they had
much higher carbon contents than other chondrites), and the **enstatite
chondrites** (named for their high abundances of the magnesium silicate
mineral, enstatite). The **Rumuruti** and **Kakangari chondrites** (both
named for the only meteorite falls of this type) may represent distinct
clans, but each contains only one recognized group.

Besides their chemical compositions, another important distinction
among chondrite clans is their oxidation states. **Oxidation** (chemical
reaction with oxygen) causes iron metal to be converted into iron oxide
or iron-bearing silicates. Increased oxidation is reflected in a change in the
relative proportions of the two most important minerals of chondrites,
olivine and pyroxene. Both of these minerals are magnesium iron silicates,
differing primarily in the ratio of silicon to magnesium plus iron. The oxidation of iron metal provides iron oxide that must be accommodated
in silicates; because the total amount of silicon is fixed, oxidation causes
pyroxene to be converted to olivine, which contains proportionately less
silicon. The least-oxidized enstatite chondrites are mostly pyroxene with
abundant iron metal, whereas moderately oxidized ordinary chondrites
have twice as much olivine as pyroxene and less metal. The oxidation state
of the Kakangari group is intermediate between that of the enstatite and
ordinary chondrites. Olivine is more abundant in the oxidized Rumuruti
chondrites, and carbonaceous chondrites, the most highly oxidized clan,
commonly contain little or no iron metal at all. Increased oxidation is
also reflected in higher iron/magnesium ratios in olivines and pyroxenes,
and their chemical analyses by **electron microprobe** provide a rapid
way to assign chondrites to their proper groups.

The groups that comprise the chondrite clans are also distinguished
by their chemical compositions and oxidation states, as well as by other

physical characteristics. In 1953, the ordinary chondrites were first subdivided, based on the ratio of metallic iron to oxidized iron, into groups with high oxidized iron (referred to as the H group) and low oxidized iron (the L group). Subsequent analyses indicated the existence of a third group with even lower oxidized iron, called the LL group. The enstatite chondrite clan also contains meteorites with differing amounts of oxidized iron, and it has now been subdivided into the EH and the EL groups. The carbonaceous chondrite clan contains a large variety of groups, each named for a type specimen: CI for Ivuna (Tanzania), CM for Mighei (Ukraine), CV for Vigarano (Italy), CO for Ornans (France), CR for Renazzo (Italy), and CK for Karoonda (Australia). Carbonaceous chondrites have compositions that most closely match that of the Sun, and the CI group is commonly taken as the best approximation of cosmic elemental abundances. The Rumuruti clan is small, consisting only of the Rumuruti (R) group, named for a 1934 fall in Kenya. Most of these meteorite groups have subtle differences in appearance, and they can sometimes be classified by knowledgeable meteoriticists on the basis of their appearance alone. There are also a few oddball chondrites, like those of the Kakangari grouplet, that do not fit into any of these clans; it has become accepted practice not to name a group until it contains at least five meteorites, although I have included the three currently recognized members of the Kakangari grouplet in this summary.

Another important distinction among chondrite groups is their oxygen isotopic compositions. Oxygen consists of three isotopes (^{16}O, ^{17}O, and ^{18}O), all stable but differing markedly in abundance (^{16}O is the most abundant). It is conventional practice to express oxygen isotopic compositions as $\delta^{17}O$ and $\delta^{18}O$, which are defined as the respective ratios of $^{17}O/^{16}O$ and $^{18}O/^{16}O$ relative to a standard, normally taken to be average ocean water. When $\delta^{17}O$ is plotted against $\delta^{18}O$, each chondrite group occupies a distinct portion of the diagram, as seen in Figure 2.5. A meteorite's oxygen isotopic composition is commonly considered proof of its proper classification.

What is the signficance of these clans and groups of chondrites with distinct chemical and isotopic compositions and oxidation states? Some researchers think that there may have been a continuum of chondrite compositions in the early solar system, with different temperatures of formation and the resulting depletion patterns of volatile elements controlled by distance from the Sun. The density and the temperature of the gas and the dust from which chondrites formed presumably decreased with increasing solar distance, resulting in more highly oxidizing conditions. In this view, enstatite and Kakangari chondrites formed

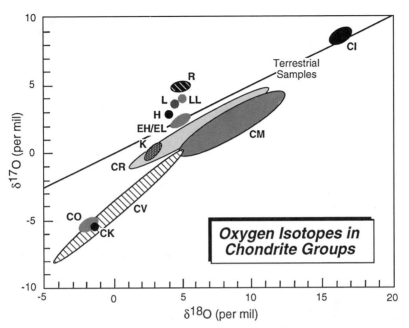

Figure 2.5: Oxygen isotopic compositions provide one means of classifying chondrites into clans and groups. In this diagram, the ratios $^{17}O/^{16}O$ and $^{18}O/^{16}O$ are expressed in δ notation, defined as parts per thousand (per mil) difference from standard mean ocean water. For reference, the diagonal line represents the locus of all terrestrial samples. All analyses are by Robert Clayton and co-workers (University of Chicago).

closest to the Sun, ordinary and Rumuruti chondrites formed at intermediate distances, and carbonaceous chondrites formed farthest away.

Chondrite Recipes

Chondrites are not well-blended consumes of ingredients, but instead are lumpy, heterogeneous aggregates of different components. Break open a chondrite and you could find that it contains, in addition to chondrules, irregularly shaped inclusions, shiny grains of metal and sulfide, and a dark, fine-grained matrix, as illustrated in Figure 2.6. Different chondrite groups contain varying proportions of these components, probably reflecting what was available in the neighborhood where each meteorite formed. Chondrites are actually a kind of cosmic sediment, composed of diverse materials with varying origins. The clumping together of these dissimilar materials is called **accretion**, and it is a very important process about which we know very little. Tiny grains probably stuck together

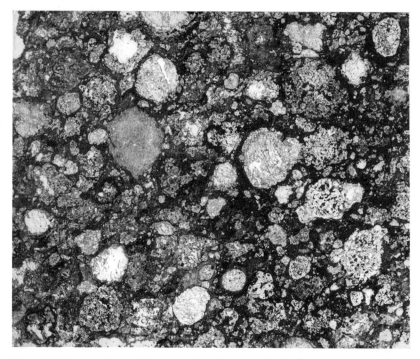

Figure 2.6: A careful inspection shows that the array of materials that comprise chondritic meteorites is more complex than it might first appear. This microscopic view of a thin section of the Vigarano (Italy) carbonaceous chondrite displays numerous rounded chondrules, irregularly shaped refractory inclusions, and opaque metal and sulfide grains, all held together by dark, fine-grained matrix material. All this diversity is contained within several square centimeters of surface area in this meteorite.

initially because of electrical charges on their surfaces, but the sticking agent for larger entities like chondrules is unknown. Each of the accreted components of chondrites probably contains a fossil record of some early solar system process or processes, but none are completely understood.

Chondrules have always been the most fascinating (and perplexing) aspect of chondrites. The German mineralogist Gustav Rose named chondrules in 1864, although references in the literature to curious globules in meteorites appeared as early as 1802. The first real breakthrough in understanding the formation of chondrules was made by Henry Sorby, who began studying chondrites at about the time Rose completed his work. Sorby was an English gentleman whose personal wealth enabled him to devote all his energies to his consuming interests in science. His major contribution was the invention of the **petrographic microscope**, which today remains one of the basic tools of geologic research. By examining paper-thin slices of rock mounted on glass slides (called thin sections), Sorby was able to elucidate features otherwise impossible to observe. After years of studying thin sections of terrestrial rocks, he finally turned his attention to chondrites. He noted that chondrules were composed of tiny crystals and glass (see Figure 2.7), similar in some respects to quickly cooled lavas. From the microscopic textures he observed, Sorby correctly concluded that these droplets had solidified from molten material, and he drew comparisons with glassy blow pipe beads and furnace slags. He also noted that these spheroids must have cooled before being incorporated into chondrites:

... melted globules with well-defined outlines could not have formed in a mass of rock pressing against them on all sides, and I therefore argue that some at least of the constituent particles of meteorites were originally detached glassy globules, like drops of a fiery rain.

Recognizing that high temperatures would be required for melting chondrules, Sorby suggested that chondrites might be either pieces of the Sun ejected in solar prominences or residual cosmic matter that had never collected into planets but that had formed when conditions at the Sun's surface extended farther out into the solar system. In the case of the first hypothesis, his views were undoubtedly colored by the concept prevailing at the time that the Sun was a solid, rocky body wrapped in incandescent gases. His second hypothesis was surprisingly close to our current concepts.

In the intervening years, a number of alternative ideas to explain the origin of chondrules have emerged. They have been suggested to be lava droplets ejected from extraterrestrial volcanoes, abraded and rounded clasts of igneous rock, liquid condensates from hot gases, splashes of melted rock from high-energy impacts, dust balls melted by lightning discharges, and clumps of solid material melted by rapid deceleration as they fell toward the Sun. This is a distressingly long list of possibilities, but it must be remembered that chondrule formation is a process with which we have no earthly experience. Chondrules do not occur in any other type of rock, except for a few isolated spheres of glass in some lunar rocks formed by impact melting. This might suggest a similar origin for meteoritic chondrules; however, impacts by themselves do not appear to be capable of making rocks composed almost entirely of closely packed chondrules.

We now know some additional facts about these enigmatic miniature marbles. For example, melting and cooling experiments that reproduce the internal textures of chondrules have shown that they were flash heated

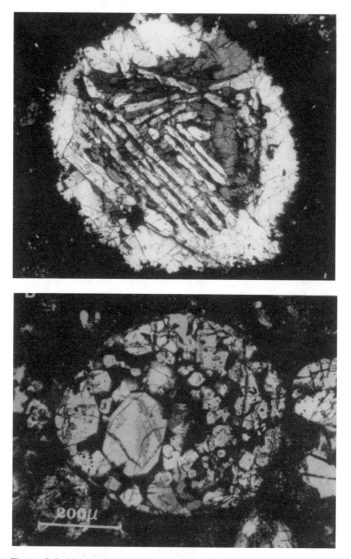

Figure 2.7: Henry Sorby's microscopic views of chondrules, seen in thin section, probably looked something like these. Shown in the upper photograph is a barred chondrule, with plates of olivine connecting to a continuous olivine rim. The porphyritic chondrule in the lower photograph contains individual crystals of olivine and pyroxene embedded in dark glass. The scale for both photographs is shown by the 200-μm bar in the bottom figure; each chondrule is approximately half a millimeter in diameter.

to above 1,500 °C and then were rapidly cooled, generally at rates of tens to perhaps a few thousand degrees per hour. These conditions seem to imply the existence of transient, localized heating events in the early solar system. Additionally, relict silicate grains in chondrules that were somehow spared from complete melting have been recognized (see Figure 2.8), suggesting that chondrules may have been repeatedly melted in multiple events. Researchers have gathered a wealth of information about the chemical and the isotopic compositions of chondrules, from which some precursor materials have been discovered. These kinds of data can

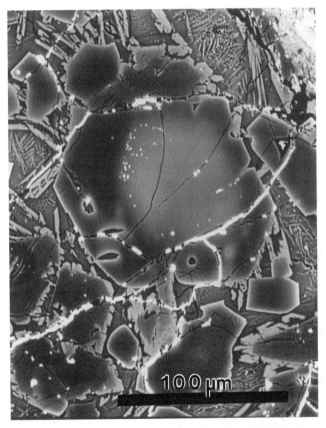

Figure 2.8: This compositional map of part of a chondrule shows a large crystal with a core of iron-rich olivine (light gray) and a rim of magnesium-rich olivine (dark gray). The tiny white dots in the core are metal grains, which do not occur in the rim. The core is a relict grain of chondrule precursor material that escaped melting, and the rim is an overgrowth that crystallized from the melt. The olivine sits in a background of chondrule glass and needles of pyroxene. Image courtesy of Rhian Jones (University of New Mexico).

be used to constrain theories of chondrule origin, but thus far no consensus has emerged, except possibly that they formed while suspended in space. Chondrules are as much a puzzle to us now as they were to Sorby.

Other interesting components of chondrites are the **refractory inclusions** (also called calcium aluminum-rich inclusions, or "CAIs"). These are especially abundant in certain groups of carbonaceous chondrites, but they occur rarely in other chondrite clans as well. Refractory inclusions take many forms (see Figure 2.9), but most appear to have some kind of concentric structure formed by layers of different minerals. The minerals in these inclusions (commonly silicates and oxides rich in calcium and aluminum, such as melilite and spinel) tend to crystallize at high temperatures, and for this reason they are thought to have been some of the first materials in chondrites to form. Some inclusions have internal textures that are similar to those of chondrules and signify that they crystallized from liquids, but others are ambiguous. Melting experiments on refractory inclusions have reproduced their minerals and textures by

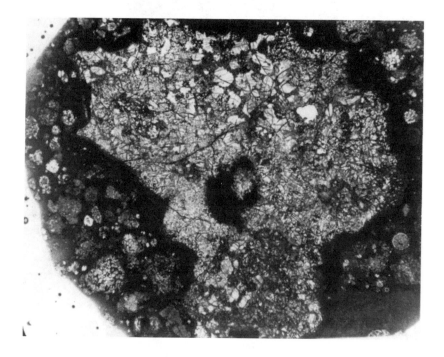

Figure 2.9: This large refractory inclusion in the Allende (Mexico) carbonaceous chondrite measures 1.7 cm across. The inclusion is viewed in thin section under a petrographic microscope. Photograph courtesy of Glenn MacPherson (Smithsonian Institution).

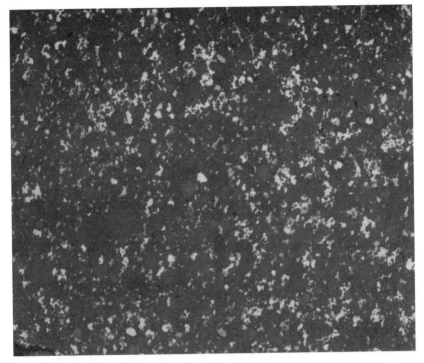

Figure 2.10: Most chondrites contain disseminated grains of metallic iron–nickel, as revealed in this polished slab of the Hedeskoga (Sweden) ordinary chondrite. Photograph courtesy of the Smithsonian Institution.

heating to temperatures as high as 1,700 °C, followed by cooling at rates of a few tens of degrees per hour, slower than most chondrule cooling rates. Some inclusions also contain tiny nuggets of platinum and other rare-metal alloys, called *Fremdlinge* (German for little strangers).

Most chondrites are speckled with small bits of highly reflective **metal** and brassy **sulfide** (see Figure 2.10). The metal consists of intergrown iron–nickel alloys with different compositions and crystal structures. Two alloys are commonly present: Taenite has a higher nickel content than kamacite, and the atoms in the former are more closely packed together. The sulfides are mostly a combination of sulfur with iron (the mineral troilite), although enstatite chondrites contain a bewildering variety of sulfides, including combinations with calcium (oldhamite), manganese (alabandite), magnesium (niningerite), zinc (sphalerite), and iron and chromium (daubreelite). Enstatite chondrites are so highly reduced that elements normally combined with oxygen are forced to bind with sulfur to produce sulfides or, in some cases, nitrogen to produce nitrides. Differences in the amounts of accreted metals partly explain the compositional distinctions among the H, L, and LL groups of the ordinary chondrite clan. Enstatite chondrites contain significantly more metal than other chondrite clans, and carbonaceous chondrites contain the least.

The fine-grained **matrix** materials that cement the chondrules and other components together consist of tiny silicate grains of olivine and pyroxene, along with minor sulfides, oxides, feldspathoids, and sometimes graphite, a form of carbon. Many chondrules are surrounded by rims of matrix material, and these coatings probably were already present when the chondrules accreted. Within the matrix are tiny but identifiable grains of interstellar dust that formed outside the solar system and were incorporated without change into chondrites. The bulk of the matrix, though, is probably matter that was reprocessed in the neighborhood of the infant Sun.

A Warm, Fuzzy View

Driving in heavy fog is an uncomfortable experience for most people because the fog limits our ability to see signs, landmarks, and other traffic. Some features in chondrites act in a similar way to obscure our ability to read the record of early solar system processes. Most chondritic meteorites have been affected to some degree by later events that have altered them from their original state; 4.56 billion years is a long time for chondritic matter to remain untouched by other happenings, even in the relative isolation of space. We now examine to what extent chondrites have survived without change. Although most chondrite researchers are concerned primarily with identifying chondrites in which the records are least altered, we now see that the secondary events experienced by many chondrites are interesting in themselves.

A nuclear reactor requires a cooling mechanism because one of the by-products of fission is heat. We have already considered the decay of radionuclides in chondrites as a means of determining their ages, and some isotopes, especially those that decay rapidly, were probably capable of generating a considerable amount of heat. Because heat can readily migrate out of small rocks in space, significant increases in temperature would occur only after chondrites had accreted into larger asteroidal bodies, sometimes called **planetesimals**. One of the most plausible radionuclides for planetesimal heating is a short-lived isotope of aluminum, ^{26}Al (its half-life is only approximately 0.7 million years). Evidence for the former existence of live ^{26}Al in chondrites is indicated by the presence of its decay product, an isotope of magnesium, located within aluminum-rich crystals that normally exclude magnesium.

Another suggested heating mechanism is the resistance to the flow of electric currents produced in planetesimals by magnetic fields induced by high-intensity solar winds. Observations of some very young stars, called T Tauri stars, suggest that they eject large quantities of matter in strong winds; however, these winds tend to be ejected from the stars' poles, rather than outward from their equators, where planetesimals are thought to have formed.

Because rocks are notoriously poor conductors of heat, the heat generated inside a chondritic planetesimal could not readily escape. Increased temperatures in the interior caused **metamorphism**, the adjustment of the minerals in chondrites to hotter conditions. Recrystallization of mineral grains in most chondrites blurred the distinctive grainy texture of chondrules and inclusions embedded in the fine-grained matrix (see Figure 2.11). In severely heated chondrites, outlines of the original chondrules may be unrecognizable. Mineralogical changes also occurred during metamorphism. Volumetrically, the most important minerals in chondrites are olivine and pyroxene, both silicates capable of wide ranges of magnesium–iron substitution (they are sometimes described as solid solutions of magnesium silicate and the iron silicate). In unmetamorphosed chondrites, the magnesium and the iron contents of these minerals are observed to vary widely, but metamorphism caused a narrowing of this range as individual grains equilibrated with each other at high temperatures. Another change was the crystallization of glass in chondrules to form feldspar. All glasses are inherently unstable and tend to form crystals spontaneously; however, the process is slow and may require some time. Antique window glass may become clouded because of partial reorganization of the atoms into very small crystals. This process is accelerated at higher temperatures, and glass is absent from metamorphosed chondrites. The metamorphic effects seen in chondrites indicate that some of them experienced temperatures approaching 1000 °C.

Various levels or grades of metamorphism have been distinguished, based on observable textural and mineralogical changes. In a seminal paper in 1967 defining these metamorphic grades (called **petrologic types**), Randy Van Schmus and John Wood devised a chondrite classification scheme that combined metamorphism with the compositional groups already discussed. Petrologic types were numbered from 1 to 6, reflecting increasing metamorphic grade, and the groups were identified by letters, such as H, L, or LL. Thus a particular chondrite can be classified by a shorthand notation; for example, H4 identifies an H ordinary chondrite of petrologic type 4. It was originally thought that type 1 chondrites were the lowest grade, but the least-metamorphosed chondrites of any group

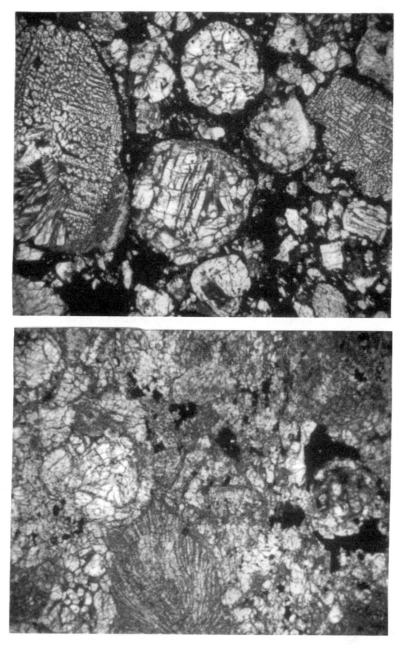

Figure 2.11: Many chondrites have experienced heating that resulted in thermal metamorphism. The most obvious effect of this process was recrystallization, the formation of new mineral grains in the solid state. These new grains are more homogeneous in their chemical compositions than the corresponding grains in unmetamorphosed chondrites. These photographs compare magnified views (each approximately 5 mms across) of a relatively unmetamorphosed (type 3) ordinary chondrite from Bishunpur (India) at the top and a severely metamorphosed (type 6) chondrite from Dhurmsala (India) at the bottom. Recrystallization of the type 6 chondrite has blurred the distinctive outlines of all but a few chondrules.

actually appear to be of type 3 (these are also called unequilibrated chondrites).

Types 1 and 2 chondrites are represented only by some groups of carbonaceous chondrites, and we now recognize that these meteorites have experienced a different kind of processing – **aqueous alteration**. Experimental constraints on the alteration of CM2 chondrites indicate that they reacted with water at temperatures below 20 °C. The more pervasively altered CI1 chondrites experienced slightly warmer temperatures, approximately 50 °C. Some researchers favor aqueous alteration in space by reactions between small solid grains and water vapor, but reactions with liquid water in meteorite parent bodies after accretion seems more likely. Wherever alteration occurred, the matrix materials in CM and CI chondrites, originally probably a mixture of fine-grained olivine, pyroxene, and smaller amounts of other minerals as seen in type 3 chondrites, were transformed into an array of hydrated minerals called phyllosilicates. These are very complex phases, commonly with layered structures similar to those of terrestrial clays. Iron oxide minerals like magnetite probably also formed at this stage. Somewhat later, fluids moved through fractures in these chondrites, precipitating veins of carbonate and sulfate minerals (such mineralized veins could have formed only in parent bodies, not in space). Most CV3, CO3, and a few type 3 ordinary chondrites experienced only minor aqueous alteration, whereas CR2, CM2, and CI1 chondrites were affected more severely (see Figure 2.12). It has also been proposed that a few CV3 chondrites suffered aqueous alteration followed by thermal metamorphism, so that their hydrous phyllosilicates were subsequently dehydrated.

The Van Schmus–Wood classification system for chondrites can be reinterpreted to take aqueous alteration into account, as illustrated by the arrows at the top of Figure 2.13. The box diagram identifies the various recognized chondrite groups, summarizes the relative proportions of petrologic types in each group, and specifies the approximate temperature range for its metamorphism or alteration. It is clear that most chondrites of the ordinary and the enstatite chondrite clans are highly metamorphosed. All currently known Rumuruti and Kakangari chondrites are of type 3, but most Rumuruti chondrites contain fragments of metamorphosed R5 or R6 material. Some groups of the carbonaceous chondrite clan (CI, CM, and CR) consist entirely of meteorites that have experienced aqueous alteration, whereas others appear to have suffered minimal alteration or metamorphism.

The chondrite classification scheme is, of course, a simplification, because each petrologic type actually represents a range, albeit limited, of metamorphic or alteration effects. Because relatively unmetamorphosed

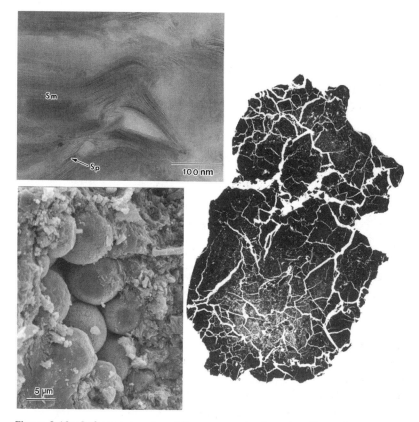

Figure 2.12: Carbonaceous chondrites commonly show the effects of aqueous alteration at low temperatures. The upper left-hand photograph is a transmission electron microscopic image of phyllosilicates in the matrix of the Ivuna (Tanzania) CI chondrite, showing grains of serpentine (Sp) and smectite (Sm). These hydrous minerals have layered structures, as revealed by the tiny, parallel lines in this image by Douglas Ming (NASA Johnson Space Center). Carbonaceous chondrites also contain magnetite formed during alteration, as shown in the image at lower left. These tiny magnetite spheres in the Haripura (Japan) CM chondrite were imaged by Marianne Hyman and her colleagues (Texas A&M University) using an electron microscope. The photograph at right shows a microscopic view of the Orgueil (France) CI chondrite. Fluids traveling in fractures deposited veins of carbonate and sulfate minerals, which appear white against the dark matrix of altered phyllosilicates.

or unaltered chondrites carry information that is least obscured by these later events, it is important to distinguish the most unequilibrated meteorites from among the type 3 chondrites. Derek Sears and his students at the University of Arkansas have developed a useful metamorphic scale for unequilibrated chondrites, based on changes in their **thermoluminescence** (normally abbreviated as TL, this technique measures the luminescence induced by a dose of radiation). During mild metamorphism, one

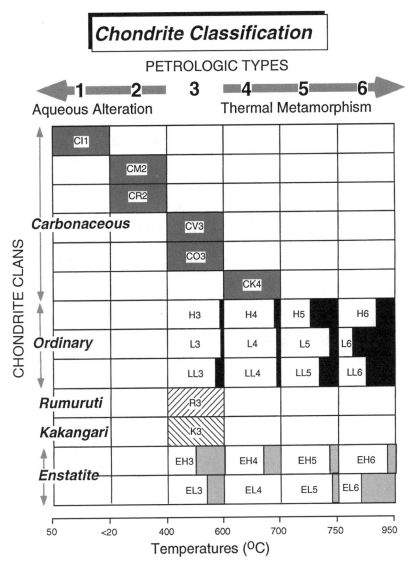

Figure 2.13: This interpretation of the commonly used classification system for chondrites consists of a grid of meteorite groups, identified by letters, and petrologic types, identified by numbers. Chondrites are sorted into groups according to their chemical and isotopic compositions and into petrologic types based on their observable properties. Each group is thought to represent meteorites derived from a single parent body, and the petrologic types are probably samples from the same body, but with distinct histories of thermal metamorphism or aqueous alteration. Arrows at the top indicate relative degrees of metamorphism and alteration, and estimated temperatures required for producing the various petrologic types are shown across the bottom of the figure. The shaded areas of each box correspond to the relative proportions of the various petrologic types for each chondrite group, showing that few chondrites are unaffected by these secondary processes.

of the first observable changes is the formation of tiny grains of feldspar from chondrule glasses. This feldspar causes a marked increase in TL sensitivity, and its measurement allows the assignment of type 3 chondrites to ranks of 3.0 to 3.9, reflecting subtle increases in the degree of metamorphism.

It is of obvious interest to know when the metamorphic and alteration events took place. The uranium–lead isotopic clock in a chondrite is set only when it cools through a certain threshold temperature (called the blocking temperature), approximately 450 °C for the phosphate mineral that contains most of the uranium. The most precise lead isotope measurements for chondrites, made by Crista Göpel and her colleagues at the University of Paris, indicate that phosphates in chondrite parent bodies reached peak metamorphic temperatures and then cooled through the blocking temperature within 60 million years of the solar system's formation. Thus metamorphic heating must have followed closely on the heels of planetesimal accretion. This short time interval is consistent with the idea that rapid heating was caused by the decay of short-lived radionuclides or electromagnetic induction. Aqueous alteration was likewise an early process. Precise measurements of radiogenic ^{53}Cr in carbonates of CI1 chondrites by Magnus Endress and co-workers at the University of Münster in Germany indicate that aqueous alteration occurred within 50 million years of the birth of the solar system, essentially the same short time period.

A Shocking Conclusion

At some point, chondrites were excavated and liberated from their parent bodies. High-velocity impacts between orbiting planetesimals can eject crushed materials from their surfaces or, in some cases, fragment such bodies. Collisions were probably common in the early solar system, so it seems likely that many meteorites might have experienced several impacts before being extracted from their parent bodies. Impact processes produce **breccias**, fragmental rocks composed of angular clasts of varying sizes and shapes (see Figure 2.14). One study of 850 ordinary chondrites by Alan Rubin and co-workers at the University of California at Los Angeles found that 5% of H, 22% of L, and 23% of LL chondrites were breccias. In most cases, chondrite breccias contain no fragments of chondrite groups other than their own, but the clasts that are present vary in petrologic type. This suggests that each parent body exhibited a range of metamorphic grades, and impacts excavated the more deeply buried (presumably highly metamorphosed) material and mixed it with unmetamorphosed surface

Figure 2.14: Impacts into the surfaces of chondrite parent bodies have fragmented the target materials, and in many cases this rubble has been subsequently compacted into coherent rocks. An example is this chondrite breccia from Nakhom Pathom (Thailand). Clearly visible angular clasts are contained within a darker, pulverized matrix formed of finely comminuted chondritic material. Photograph courtesy of the Smithsonian Institution.

material. Breccias composed of carbonaceous, enstatite, and Rumuruti chondrites are also abundant. Carbonaceous chondrite clasts, especially of the CM group, are more commonly included as foreign components (**xenoliths**) of other chondrite group breccias than are other meteorites, so this material must have been flung far and wide.

Many breccias have been affected by **shock metamorphism**. The collision of two incoming projectiles traveling at relative velocities of several kilometers per second instantaneously produces immense pressures, although they persist for only a few seconds. Values of at least 75 GPa (approximately 750,000 times the Earth's surface atmospheric pressure) have been documented in chondrites, and the release of pressure was accompanied by significant heating. Some impacted meteorites have been melted, and others have experienced severe deformation of their crystal structures. Shocked minerals are sometimes transformed into more densely packed forms, just as they are in the target rocks of terrestrial impact craters. Nearly 17% of chondrite falls show the clearly discernible effects of shock metamorphism. Dieter Stöffler at the University of Münster and Klaus Keil and Edward Scott of the University of Hawaii devised a scheme for classifying chondrites by shock intensity, in ascending order from S1 to S6, analogous to the scale for degree of thermal metamorphism.

Such impact processes have affected chondrites throughout their history, but they were probably most potent in the early solar system before the planets swept up much of the smaller orbiting matter. Most impacts appear to have taken place after thermal metamorphism (which occurred in the first 60 million years or so after chondrite formation), because chondritic breccias typically contain clasts of different metamorphic grades. However, some impacts must have occurred while metamorphism and aqueous alteration were still happening. Shock metamorphism can sometimes be dated by radioactive methods. The decay products of a few radionuclides are gases, such as an isotope of argon, ^{40}Ar, which forms from the decay of ^{40}K, an isotope of potassium. Shock metamorphism can disturb a meteorite sufficiently for argon gas to escape. As the meteorite subsequently cools through its argon-blocking temperature, ^{40}Ar again begins to accumulate and the radioactive clock is reset. The age derived from this kind of isotope system, called a **gas-retention age**, represents the most recent thermal disturbance that affected the meteorite. For breccias, this disturbance was shock metamorphism. Measured gas-retention ages for ordinary chondrites range from 4.1 billion to less than 0.1 billion years, reflecting random impact events.

All these secondary events affecting chondrites provide an interesting picture of chondrite parent body evolution. However, the record of early solar system processes in chondrites can be obscured by these overprints. It may seem surprising that we can read any primary record through this haze of thermal metamorphism, aqueous alteration, brecciation, and shock effects. It is possible, through careful observations and measurements, to sort out these secondary events in meteorites and in some cases to compensate for their effects. However, most of what we know about early solar system processes comes from the chondrites that have suffered the least modification, the chondrites of petrologic type 3.

Micrometeorites

Micrometeorites, with sizes measured in fractions of millimeters, are often melted during transit through the atmosphere and thus retain little information useful for their classification. IDPs, however, are collected high in the stratosphere, thus minimizing their heating. IDPs are almost always chondritic. They come in several varieties, commonly called chondritic smooth (CS) and chondritic porous (CP) particles to distinguish their appearance. Classifying small IDPs is challenging, because they may not be representative samples and because the minerals they contain are so tiny. Another potential pitfall is that some IDPs are at least modestly heated during atmospheric passage, producing a surficial layer of oxidized iron minerals. Nevertheless, careful observations with electron microscopes have allowed some of these miniature particles to be classified.

A handful of well-studied CS dust particles have been shown to contain phyllosilicates and carbonates. Their chemical compositions are broadly chondritic, but they are slightly depleted in elements like calcium and magnesium that can be mobilized during aqueous alteration. Hence they appear to be dust motes of known groups of carbonaceous chondrites, such as CI and CM chondrites. In contrast, CP particles consist mostly of anhydrous minerals like olivine and pyroxene. Although chondritic in composition, many of these IDPs contain abundant carbon, sometimes greatly exceeding the proportion of carbon in chondritic meteorites. The CP dust particles do not appear to conform to any known meteorite group.

Reading the Record in Chondrites

In order to unravel the secrets in unaltered type 3 chondrites, we must digress a bit to discuss current concepts of how the solar system formed. A gravitationally contracting cloud of interstellar gas and dust is thought to have formed a central mass concentration (which in the case of our solar system would ultimately become the Sun), and its rotation flattened the outlying material into a disk, called the **solar nebula**. Such nebular disks around infant stars in distant molecular clouds have been observed by the *Hubble Space Telescope*. Contraction would have caused heating of the solar nebula, and some calculations suggest that the inner parts of the disk were sufficiently hot to have vaporized all the dust. On cooling, this vapor would have partly recondensed as new grains. In the Earth's atmosphere, water vapor condenses as liquid (that is, raindrops) but, at the low pressures thought to occur in the nebula, gas would have condensed to solids. **Thermodynamic calculations** have been used to predict the order of appearance of minerals that condensed from a cooling nebula gas of cosmic composition. Some of these minerals condensed directly, but the **condensation sequence** is complicated by reactions of already-condensed minerals with nebular gases to form other minerals as cooling proceeds (see Figure 2.15).

All the minerals in the theoretical condensation sequence actually occur in chondrites. Those that are predicted to have condensed or formed by reactions at the highest temperatures – corundum, hibonite, perovskite, melilite, spinel, diopside – make up the bulk of refractory inclusions in carbonaceous chondrites. Similar refractory inclusions also occur in ordinary chondrites, although they are relatively uncommon. The minerals in the middle of the condensation sequence – olivine, pyroxene, metallic iron, plagioclase – occur as common constituents of chondrules. The low-temperature end of the sequence is represented by minerals in the matrix – iron-rich olivine and pyroxene, magnetite, troilite, and so forth. Many condensates were subsequently melted and then recrystallized to form inclusions and chondrules before they were incorporated into the meteorites.

The calculated condensation sequence is very sensitive to the amount of oxygen present, and a different sequence is predicted for more reducing gases. For a vapor that has less oxygen and a greater proportion of carbon than for cosmic abundances, the minerals oldhamite, osbornite, niningerite, and alabandite take the place of hibonite, perovskite, melilite, and spinel as the earliest condensing phases. Enstatite chondrites contain oldhamite and the other minerals predicted to condense from less oxidizing gases (although, like the refractory inclusions in other chondrite groups, some of these minerals were apparently melted and recrystallized before being accreted into meteorites). The existence of enstatite chondrites may thus imply that different parts of the solar nebula had varying compositions when condensation occurred.

The high-temperature condensation of the minerals in refractory inclusions requires that they must be among the earliest-formed materials

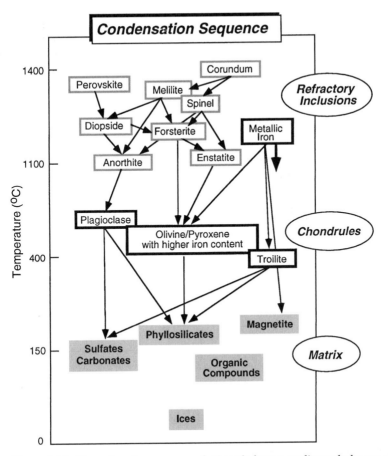

Figure 2.15: The order of appearance of minerals from a cooling nebular gas of cosmic composition has been predicted from theoretical calculations and is schematically summarized in this figure. Minerals like perovskite, melilite, and corundum that form at high temperatures condense directly from the gas, whereas those that form at lower temperatures result from reactions of the gas with previously condensed minerals (reactions are indicated by arrows). The condensation of metallic iron is suppressed, relative to forsterite and enstatite, at lower pressures (indicated by a large arrow). Although this condensation sequence is probably an oversimplified view of the formation of solid matter in the solar system, it does predict the occurrence of minerals that comprise refractory inclusions, chondrules, and matrix. The calculated condensation sequence is based on the work of Larry Grossman (University of Chicago).

in the solar system. There is evidence to support that idea. The extinct radionuclide ^{26}Al, already mentioned as a possible heat source for chondrite metamorphism, decayed to ^{26}Mg very rapidly. The presence of excess ^{26}Mg has been verified in refractory inclusions from chondrites, and its incorporation into aluminum-bearing minerals that are normally magnesium free implies that the ^{26}Al was incorporated live. Because the refractory inclusions incorporated more live ^{26}Al than other chondritic materials, they are thought to have been the earliest-formed components of chondrites. Other isotopic clocks, such as the rubidium strontium system, applied to refractory inclusions confirm their antiquity.

The separation of one chemical component from another is called **fractionation**. Incorporation of varying amounts of refractory inclusions probably explains the observed fractionations of highly refractory elements among chondrite groups. The fractionation of highly volatile elements, at the other end of the condensation sequence, also relates to the formation of solids from vapor, but in this case condensation was far from complete. Volatile elements are depleted, relative to cosmic abundances, in virtually all chondrite groups and clans, but not to the same degree. Elements with similar volatility have similar concentrations in chondrites, but those that are more volatile tend to be less abundant. Figure 2.16 compares volatile depletions in CM carbonaceous chondrites and unequilibrated ordinary chondrites, relative to cosmic (CI) abundances of these elements. In this figure, the elements are ordered by increasing volatility from left to right, and the tilts of the data points demonstrate that the elements that are the most volatile are also the most depleted.

Not all chemical fractionations observed in chondrites, however, can be explained by volatility. An important fractionation of this type is the separation of metal from silicate. As noted previously, differences in chondrite groups are partly attributable to the incorporation of different amounts of iron–nickel metal, as well as differences in oxidation state. All chondrite groups (except CI) are depleted in iron relative to the cosmic abundance, as illustrated in Figure 2.17. In this plot of metallic iron versus oxidized iron, reduction (the opposite of oxidation) moves the cosmic abundance composition (CI) diagonally, but all chondrite groups fall well below this line, indicating loss of iron. Concurrent depletion of nickel and other metal-loving elements with iron demonstrates that the lost iron was metallic rather than oxidized. The mechanism for metal fractionation is not well understood, but it appears that metal grains may have approximately the same mass, on average, as do chondrules in the same meteorite. Sorting of chondrules and metal grains, presumably by some physical process in the nebula that segregated objects of similar mass, might explain the observed metal-silicate fractionations in chondrites.

Another aspect of the early solar system that is revealed in chondrites is the degree to which mixing of different nebular regions occurred. Although clasts of different meteorite types, especially carbonaceous

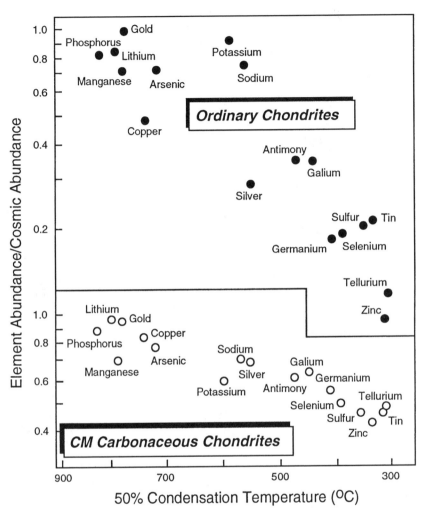

Figure 2.16: The element abundances of various chondrite groups show depletion patterns related to element volatility. This diagram compares element abundances in ordinary chondrites and CM carbonaceous chondrites, both relative to cosmic (CI) abundances. The degrees of depletion, reflected in the slopes of these trends, are distinct. Volatility (expressed as the temperature at which 50% of the element would have condensed) increases from left to right in the diagram. The chemical analyses were performed by John Wasson (University of California at Los Angeles).

chondrites, occur within breccias of other chondrite groups, this evidence of mixing postdates the formation of planetesimals. To gain information on nebular mixing, we must turn to isotopes. The three stable isotopes of oxygen can be separated from each other by various processes, but the fractionation is always proportional to the differences in the masses of the

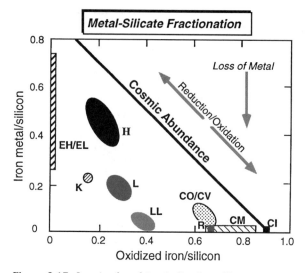

Figure 2.17: Iron in chondrites is distributed between coexisting metal and oxidized forms (mostly silicates). The cosmic abundance of iron relative to silicon, regardless of its oxidation state, is indicated by the diagonal line through the CI chondrite composition. All other chondrite groups fall below this line, reflecting loss of iron metal before their accretion.

isotopes. Thus any event that increases ^{17}O relative to ^{16}O (a difference of one mass unit) by a small amount will increase ^{18}O relative to ^{16}O (a difference of two mass units) by twice that amount. If we plot the ratios of $^{17}O/^{16}O$ versus $^{18}O/^{16}O$ (expressed as $\delta^{17}O$ and $\delta^{18}O$), mass-fractionated isotopes will be smeared along a straight line with a slope of $+1/2$, as shown in Figure 2.18. Virtually all terrestrial materials, whether rocks, air, or living organisms, fall along this mass-dependent fractionation line. However, the oxygen isotopic compositions of refractory inclusions and chondrules in carbonaceous chondrites fall along a line with a different slope, as shown in the same figure. How could these isotopic variations arise? This second line extrapolates to nearly pure ^{16}O at the origin of the figure, and thus compositions along the line could be produced when pure ^{16}O is mixed with normal solar system oxygen, located perhaps at the intercept of the mixing line with the terrestrial mass-dependent fractiona-tion line. The mystery then becomes, where did the pure ^{16}O come from, and how was it mixed with normal solar system material? To produce pure ^{16}O, we require some kind of nuclear process, and the only reason-able possibilitiy seems to be a **supernova**, the explosion of a massive star. During such an event, any ^{17}O or ^{18}O in the star theoretically would be destroyed by the intense nuclear reactions, and only ^{16}O would survive.

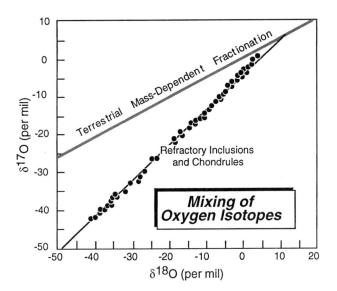

Figure 2.18: Oxygen isotopes in the refractory inclusions and chondrules of carbonaceous chondrites fall along a mixing line between pure ^{16}O and normal solar system oxygen, possibly plotting near the intersection of the mixing line with the terrestrial mass-dependent fractionation line. The steep slope of the mixing line indicates that this variation cannot be explained by the fractionation processes that occur on Earth. The δ notation used to describe the oxygen isotopic composition is defined in the caption for Figure 2.5. All analyses were made by Robert Clayton and co-workers (University of Chicago).

Grains formed from supernova debris and containing pure ^{16}O could then be ejected into the interstellar medium and eventually wander into the solar nebula. When these are mixed in various proportions with other, normal solar system condensates and melt droplets, their bulk isotopic compositions would fall along the observed isotopic mixing line. These isotopic anomalies demand that nebula temperatures never exceed that required for vaporizing all the ^{16}O-laden dust.

Carbonaceous Matter

An important chemical component of chondrites that has not yet been introduced is **organic matter**. These materials are complex and sometimes rather large molecules composed mostly of carbon, with hydrogen, oxygen, nitrogen, or sulfur. Such materials occur in all chondrites but have been studied extensively only in carbonaceous chondrites. It is important to mention contamination before we assess the results of organic analyses, as the quantity of indigenous hydrocarbons is equal in some cases to that transferred to the meteorite by just a few fingerprints. Many chondrites have been handled or stored under less than sterile conditions in museum collections or could have been exposed to terrestrial weathering before collection. Therefore analyses of organic matter in meteorites must be viewed with caution.

The discovery of small (millimeter or smaller) complex clumps of organic material in chondrites spurred controversy some years ago. These so-called organized elements were thought by some to be biological in origin and were touted as evidence that primitive life forms were carried to Earth by meteorites. However, these were contamination products, such as airborne pollen grains implanted in the meteorites while on our planet. At least one example in the Orgueil (France) meteorite appears to have been an elaborate hoax.

Most of the organic matter in chondrites, however, does appear to be indigenous. This material is a complex mixture of straight to slightly branching hydrocarbon chains (alkanes), rings (aromatic hydrocarbons), and carboxylic and amino acids. The organic matter in carbonaceous chondrites apparently formed within meteorite parent bodies, probably at the same time that they experienced aqueous alteration. However, the forerunners of these compounds were simpler molecules formed in interstellar space. The hydrogen they contain is isotopically heavy (with large amounts of D, deuterium), which is characteristic of organic molecules forged in interstellar clouds. Therefore the organic fraction of chondrites, much like that of the inorganic materials, is an amalgam of interstellar matter that has, for the most part, undergone processing in the solar nebula and within planetesimals.

IDPs also contain organic matter, but because they are so small their carbonaceous matter is not as well characterized as in chondrites. A few aromatic hydrocarbons have been identified, and some IDPs contain amorphous carbon.

A Key for Decoding Secrets

Eighteenth-century scholars had little success in deciphering hieroglyphs carved on ancient Egyptian monuments until French engineers found a tablet in 1799 while repairing a fort at the mouth of the Nile River near the town of Rosetta. On this tablet, the now-famous Rosetta Stone, are hieroglyphs and their translations in Coptic and Greek. This tablet unlocked the history of an entire ancient culture.

In a very real sense, chondritic meteorites are Rosetta Stones that have enabled us to decipher the earliest history of the solar system. The average

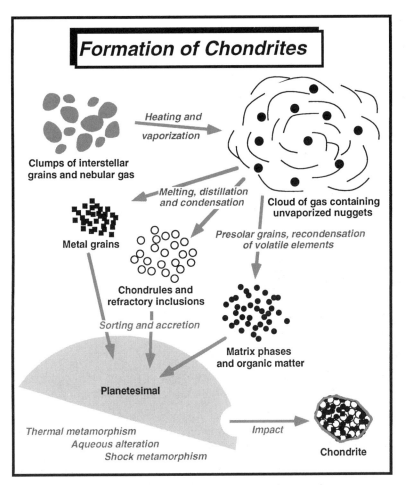

Figure 2.19: The formation of chondrites involved a series of complex steps, as illustrated in this figure.

chemical composition of the solar system and the timing of its birth are measureable only in chondrites. Chondrules, refractory inclusions, and organic compounds in these meteorites represent interstellar matter that was reprocessed in the early solar system (see Figure 2.19). The diverse mineralogical components of chondrites provide an unparalleled record of nebular processes like condensation, chemical fractionation, and mixing, despite the overprints of later events that altered chondrites to varying degrees. Organic compounds in these meteorites may have been the basic building blocks for life. Chondrites are truly unique specimens, a jumbled assortment of nebular dust and organic muck that holds the key to unlocking the secrets of our solar system's origin.

Suggested Readings

Most of these readings are at a more technical level than this book, because elementary descriptions of chondrites have not commonly been published. However, they are all interesting contributions that will provide rich sources of information on chondrites for the serious reader.

GENERAL

Wasson J. T. (1985) *Meteorites: Their Record of Early Solar-System History*, Freeman, San Francisco, CA.

Chapter II of this book provides a thorough discussion of chondrite composition and taxonomy, and Chapter III discusses radiometric ages.

Kerridge J. F. and Matthews M. S., eds. (1988). *Meteorites and the Early Solar System*, University of Arizona, Tucson, AZ.

Part 3.3, by H. Y. McSween, Jr., D. W. G. Sears, and R. T. Dodd, describes the thermal metamorphism of chondrites; Part 3.4, by M. Zolensky and H. Y. McSween, Jr., introduces aqueous alteration; and Part 3.6, by D. Stöffler, A. Bischoff, V. Buchwald, and A. E. Rubin, provides an overview of shock metamorphism in chondrites.

CHONDRITE AND IDP CLASSIFICATION

Van Schmus W. R. and Wood J. A. (1967). A chemical-petrologic classification for the chondritic meteorites. Geochim. Cosmochim. Acta **31**, 747–765.

A classic technical paper that first formulated the currently used, but now somewhat modified, classification system for chondrites.

Sears D. W. G., Grossman J. N., Melcher C. L., Ross L. M., and Mills A. A. (1980). Measuring metamorphic history of unequilibrated ordinary chondrites. Nature (London) **287**, 791–795.

This technical paper describes the use of thermoluminescence in fine-tuning the classification of relatively unmetamorphosed type 3 chondrites.

Stöffler D., Keil K., and Scott E. R. D. (1991). Shock metamorphism of ordinary chondrites. Geochim. Cosmochim. Acta **55**, 3845–3867.

This paper constructs a scale for classifying chondrites according to the intensity of shock metamorphism.

Schramm L. S., Brownlee D. E., and Wheelock M. M. (1989). Major element composition of stratospheric micrometeorites. Meteoritics **24**, 99–112.

A technical paper outlining the distinctions between chondritic porous and smooth IDPs.

METAMORPHISM, AQUEOUS ALTERATION, AND SHOCK EFFECTS

McSween H. Y. Jr. and Labotka T. C. (1993). Oxidation during metamorphism of the ordinary chondrites. Geochim. Cosmochim. Acta **57**, 1105–1114.

A technical paper documenting progressive changes in mineralogy during the thermal metamorphism of chondrites.

Buseck P. R. and Hua X. (1993). Matrices of carbonaceous chondrite meteorites. Ann. Rev. Earth Planet. Sci. **21**, 225–305.

The mineralogical changes that accompany aqueous alteration of chondrites are identified in this review.

Bischoff A. and Stöffler D. (1992). Shock metamorphism as a fundamental process in the evolution of planetary bodies: Information from meteorites. Eur. J. Mineral. **4**, 707–755.

The various shock effects seen in meteorites are described in detail.

CHEMICAL COMPOSITION

Anders E. and Grevesse N. (1989). Abundances of the elements: meteoritic and solar. Geochim. Cosmochim. Acta **53**, 197–214.

An exhaustive, and somewhat exhausting paper describing the most current cosmic abundance table of the elements, based primarily on chondritic meteorites.

Clayton R. N. (1993). Oxygen isotopes in meteorites. Ann. Rev. Earth Planet. Sci. **21**, 115–149.

This author nicely summarizes decades of painstaking oxygen isotope analyses of chondrites and their utility in classifying meteorites and in interpreting nebular and parent body processes.

Cronin J. R. (1989). Origin of organic compounds in carbonaceous chondrites. Adv. Space Res. **9**, 54–64.

This paper describes current ideas about the formation of organic matter in meteorites.

AGES

Minster J. F., Birck J. L., and Allègre C. J. (1982). Absolute age for formation of chondrites studied by the ^{87}Rb–^{87}Sr method. Nature (London) **300**, 414–419.

An excellent summary of the chronology of chondritic meteorites, based on analysis of radiogenic isotopes.

Bogard D. D. (1995). Impact ages of meteorites: a synthesis. Meteoritics Planet. Sci. **30**, 244–268.

This thoughtful summary of shock ages, determined primarily by ^{39}Ar–^{40}Ar isotope analyses, focuses on the impact histories of chondrites.

NEBULAR PROCESSES RECORDED IN CHONDRITES

Kerridge J. F. (1993). What can meteorites tell us about nebular conditions and processes during planetesimal accretion? Icarus **106**, 135–150.

This paper presents an engrossing summary of meteoritic evidence for understanding nebular physical and chemical conditions and processes.

Metzler K., Bischoff A., and Stöffler D. (1992). Accretionary dust mantles in CM chondrites: evidence for solar nebula processes. Geochim. Cosmochim. Acta **56**, 2873–2897.

Interesting observations of rims on chondrules that bear on the accretionary process.

3 Chondrite Parent Bodies

The primitive nature of chondrites demands that they come from objects that have somehow escaped severe geologic processing. This is most readily understood if their parent bodies are small. In this chapter we explore the likelihood that chondrites and chondritic micrometeorites (that is, IDPs) are derived from the smaller objects in the solar system, asteroids and comets. The major observational distinction between these two kinds of bodies is that comets are highly luminous objects and asteroids are not. The small sizes of both kinds of objects qualify them as cosmic junk. After the initial discovery of asteroids, there was a flurry of astronomical activity to catalog these minor planets. However, by the 1950s, interest had waned to such a degree that major observatories considered it a waste of time to study such "vermin of the skies." This denigration of status occurred because until just a few years ago asteroids were considered superfluous material, mere remnants of the process of planet formation. However, new observational techniques and spacecraft missions, as well as meteorite studies that have revealed the importance of these objects for understanding the early solar system, have rekindled interest in asteroids and comets.

A Connection to the Asteroid Belt

In 1959, an H5 ordinary chondrite hit the Earth at Pribram, near Prague in the Czech Republic. What made this fall unusual was that it was accidentally photographed by several cameras originally set up to track artificial satellites. Because accurate **orbits** for meteorites can be ascertained only from precise observations made simultaneously at two or more locations, Pribram became the first recovered chondrite whose orbit before Earth capture could be determined.

Partly as a consequence of the inadvertent Pribram experiment, in 1964 a network of cameras was set up in the United States by the Smithsonian Astrophysical Observatory to photograph the trajectories of any meteoroids of sufficient size to reach the ground intact (see Figure 3.1).

79

Figure 3.1: This fireball was photographed by the Prairie Network station at Hominy, Oklahoma, in early 1970. The picture has been tilted so that the ground is at a 45° angle in the lower right-hand corner. Spaces between luminous segments of the fireball's trajectory are caused by a chopping shutter used for timing and velocity determination. The faint lines in the background are star trails resulting from the Earth's rotation during the 3-hour exposure. These photographic results were used to pinpoint the fall location of this meteorite, which was recovered a few days later near Lost City, Oklahoma. Its preatmospheric orbit was calculated from sightings by a number of network stations. Photograph courtesy of the Smithsonian Astrophysical Observatory.

This Prairie Network consisted of sixteen stations disposed over a circular area approximately 500 km in radius and centered in southeastern Nebraska. The system functioned for approximately a decade. Although the orbits of hundreds of fireballs were recorded during the network's lifetime, only one meteorite with a recorded orbit was actually recovered. A similar network in Canada allowed the collection of an additional meteorite and the determination of its orbit. Both recovered meteorites, Lost City (Oklahoma) and Innisfree (Alberta), were ordinary chondrites. This was not really surprising, considering the frequency of ordinary chondrite falls. The orbits of several other ordinary chondrites, Farmington (Kansas) and Dhajala (India), have been calculated from independent observations of the falls by eyewitnesses. Their trajectories are not as accurately assessed as those determined from photographic networks, but they do enlarge this rather limited database.

The orbits for all five chondrites are shown in Figure 3.2. Each meteorite's path is elliptical and crosses inside the Earth's orbit, as it must if the chondrite is to fall on our planet. But the really interesting feature of each of these orbits is that its **aphelion** (the most distant point in its orbit), indicated in the figure by a dot, lies between the orbits of Mars and Jupiter. This is the position of the **asteroid belt**, located primarily between 2.1 and 3.3 AU (one AU, or **astronomical unit**, corresponds to the mean distance between the Sun and the Earth, approximately 150 million km). Not all asteroids are found within the main belt, however. Several other discrete populations are recognized, including the Hungarias at 1.8–1.9 AU, the Cybeles at 3.3–3.5 AU, the Hildas at 3.9–4.0 AU, and the Trojans at 5.1–5.3 AU. The Trojans actually travel the same orbital path as Jupiter, but are clustered within two stable regions located 60° ahead of and behind the giant planet.

Within the asteroid belt and its outliers are approximately 5,000 asteroids of sufficient size that their orbits are well characterized, another 13,000 smaller bodies that are known but whose orbits have not been determined, and perhaps a million asteroids altogether. As we noted in Chapter 1, asteroids without determined orbits are not assigned catalog numbers and names; instead, each is marked by its year of discovery and two letters that refer to the date of its first observation (as in 1989 PB, which later was renamed 4769 Castalia once its orbit was determined). The largest asteroid is 1 Ceres, with a diameter of more than 900 km. Despite the large number of asteroids and the fact that several hundred of them have diameters greater than 100 km, all of them taken together have only a fraction of the mass of the Moon. For most people, the term asteroid belt conjures up visions of a maze of closely spaced chunks of rock stretching as far as the eye can see, probably because of the

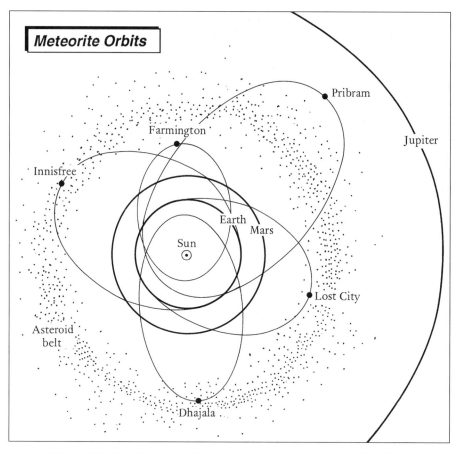

Figure 3.2: The calculated orbits of recovered meteorites provide information on the sources of these objects. Five recovered ordinary chondrites had highly elliptical orbits. All of these had aphelia, the approximate locations of which are illustrated by small dots in this figure, within or near the asteroid belt between Mars and Jupiter. This suggests that chondrites may be fragments of asteroids. The orbits are drawn to scale, but their orientations are chosen for clarity of illustration.

navigational hazard they have posed in so many science fiction movies. In reality, on their way to Jupiter and beyond, NASA spacecraft have traversed the asteroid belt blind without a scratch. Although it was formerly argued that all the asteroids were once assembled into one now-disrupted planet, modern calculations indicate that the swarm of asteroids could never have accreted into such a large body because of the perturbing effects of the planet Jupiter. The massive gravitational field of this giant neighbor would have ripped apart any larger planet within its sphere of influence as quickly as it formed.

The fact that the aphelia of chondrite orbits lie within the asteroid belt suggests that these meteorites are pieces of asteroids. Such a link was first postulated in 1805 by the German astronomer Heinrich Olbers, but of course without the supporting evidence of meteorite orbital calculations. The NEOs have highly elliptical trajectories that cross into the inner solar system, just like the orbits of the recovered chondrites. Although larger than meteorites, NEOs likewise are relatively small objects with short lifetimes in space. The NEOs were liberated by impacts from main-belt asteroids, probably along with some smaller chunks that became meteorites. Alternatively, some NEOs are probably the immediate predecessor parent bodies for chondrites, the latter released when orbiting NEOs periodically collided.

Cosmic Snowballs

Passages of comets through the inner solar system are among the most spectacular astronomical events that can be seen with the naked eye. But such isolated cometary occurrences are but the tip of the iceberg, so to speak. Most comets belong to a vast swarm, called the **Oort cloud** after the Dutch astronomer who proposed its existence, that surrounds the solar system. This comet reservoir defines a spherical volume extending out to as far as 100,000 AU; in contrast, the planets all lie inside 40 AU and within a flat plane. The objects in the Oort cloud probably number in the billions. Current models suggest that many comets may have originally formed in the vicinity of Jupiter, Saturn, Uranus, and Neptune, but early on they were scattered to the outer fringes of the solar system by gravitational encounters with these giant planets.

The denizens of the Oort cloud drift about the Sun until gravitationally influenced by the nearer stars. Such perturbations may occasionally send comets hurtling into the inner solar system. Such bodies are called long-period comets, so named because of the great intervals between successive arrivals. However, another source may provide more of the comets that traverse the inner planet region. In 1951 Gerard Kuiper of the University of Arizona proposed that there should be a remnant population of cometary bodies beyond the orbit of Neptune, but lying well inside the innermost Oort cloud. This proposal was not taken seriously until the 1980s, when calculations showed that a comet belt beyond Neptune was the most plausible source for short-period comets that make moderately frequent passes into the inner solar system. In recognition of Kuiper's insight, this reservoir was labeled the **Kuiper belt**.

In 1992, the first Kuiper belt object was discovered, and more than 30 objects with diameters greater than 100 km have now been identified. On the basis of discovery statistics, many thousands of similar objects are thought to populate this belt. A handful of similar objects, called **Centaurs**, also occur between the orbits of Saturn and Neptune (see Figure 3.3). The best-studied Centaurs, 2060 Chiron and 5145 Pholus, are similar in size to the known Kuiper belt objects. Their peculiar orbits are not stable over the lifetime of the solar system (because of interactions with Saturn), leading to the suggestion that they were once members of the Kuiper belt population. The Centaurs appear to be cometlike objects (Chiron even has a luminous shroud) with asteroidlike orbits. The

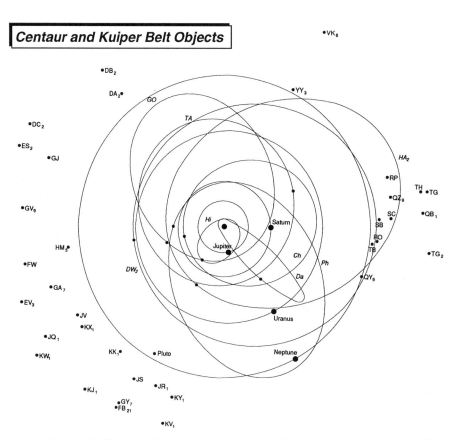

Figure 3.3: The orbits (where determined) and heliocentric distances for known Centaur and Kuiper belt objects all lie outside the orbit of Jupiter. 2060 Chiron is indicated by Ch and 5145 Pholus by Ph. These bodies are comets, some of which could potentially provide samples (mostly in the form of IDPs) to the inner solar system. Data compiled by the Minor Planet Center.

discovery in 1978 of Pluto's satellite Charon, as well as indications that both Pluto and Charon may be compositionally similar to comets, have led to consideration that these bodies also may once have been large Kuiper belt objects. Triton, a moon of Neptune with a highly unusual orbit that suggests it was captured, may represent another such object.

Comets take their name from the Greek word *kome* (hair), a reference to the spectacular, streaming tails that point away from the Sun because of interaction with the solar wind. The head of the comet is shrouded in an incandescent halo, called the **coma**. Study of the coma and the tails provides information on the composition of comets. These luminous features are composed of fragments of gaseous molecules, called free radicals, and minute particles of dust. The gases were originally amorphous (structureless) ices – mostly water, but also methane, ammonia, and other compounds – before being volatilized and split apart by solar radiation. The dust particles – silicates, oxides, and sulfides, in particles as fine as cigarette smoke – accreted with the ices as the comets formed. Also present are organic compounds of carbon, hydrogen, oxygen, and nitrogen (commonly called CHON particles, after the chemical symbol for each element). The recognition that comets are mixtures of ices and rocky particles has led to their description as dirty snowballs, although that term may overstate the proportion of ices to dust.

The properties of comets are observed to change perceptibly with continued exposure to the Sun, and the dirty snowball model may offer an explanation. Long-period comets, newly arrived from the Oort cloud, expel large quantities of gas and dust. In contrast, some short-period comets, after completing a number of trips around the Sun, appear to have burnt out, as much of the icy material capable of being volatilized has been expended. A few comets have also been observed to fragment or disintegrate on close approach to the Sun or after multiple passes. The orbits of short-period comets are very similar to those of NEOs, leading to the suggestion that some NEOs may be the devolitalized nuclei of spent comets. At least one NEO, 4015 Wilson–Harrington, has exhibited sporadic outbursts of cometary behavior. Although the observable properties of most asteroids and comets are usually distinct, the demarcation between them has become blurred with the recognition that comets may evolve into objects that closely resemble asteroids.

The Shape of Things to Come

The sizes, shapes, and spin properties of asteroids and comet nuclei can provide useful constraints on their formation and evolutionary histories, but it is usually difficult to obtain such information. Asteroid actually means starlike and, viewed through a telescope, these planetesimals are merely point sources of light. Even the largest known asteroid has a disk less than a second of arc across, which means that it is similar in appearance to many stars (although some larger asteroids can now be resolved with the *Hubble Space Telescope*). Nevertheless, the physical properties of a number of asteroids, and to a much lesser extent comets, have been determined with a variety of ground-based observational techniques.

Some asteroids exhibit changes in brightness that are due to rotation of their irregular shapes about the spin axes. The changing cross-sectional area reflects different amounts of sunlight back to observers on the Earth, and the resulting asteroid **light curve** can be modeled to estimate its shape and rate of rotation. In some cases, the variation in the light intensity can be dramatic; for example, the reflectivity of 433 Eros varies by more than a factor of 4 as it rotates. Asteroid rotation rates are typically a few hours to a few days.

Another technique used in the remote sensing of asteroids is **radiometry**, a type of heat sensing. Thermal vibrations are in the infrared region of the spectrum, and infrared measurements are used to detect heat escaping from poorly insulated buildings or in military applications to spot moving vehicles in complete darkness. A variation of this method also permits the determination of the sizes of asteroids, if some assumptions can be made about the composition of their surfaces (as discussed below). Comparison of the thermal brightness with the brightness at visual wavelengths provides a means of calculating asteroid diameters to an accuracy of approximately 10%.

Sometimes an asteroid may pass in front of a star, causing its shadow to pass over the Earth's surface. The measured silhouette provides another way to determine an asteroid's size and shape. One such experiment was performed to assess the dimensions of 433 Eros during a close approach in 1975. According to calculations completed just hours before the event, Eros was to eclipse a star in the constellation Gemini. Teams of observers rushed to preselected locations in Connecticut and recorded the duration of the **stellar occultation** at various positions along the path of the moving shadow. These observations made it possible to determine that this asteroid is a slab measuring approximately 7 km × 19 km × 30 km, tumbling end over end every five hours (see Figure 3.4).

Radar measurements give precise information on a body's size, shape, and spin rate. An intriguing result for asteroid 4769 Castalia was obtained by Steven Ostro and his colleagues at NASA's Jet Propulsion Laboratory. Their radar imagery indicates that Castalia actually consists of two discrete lobes. The most plausible explanation for Castalia's odd shape seems to be

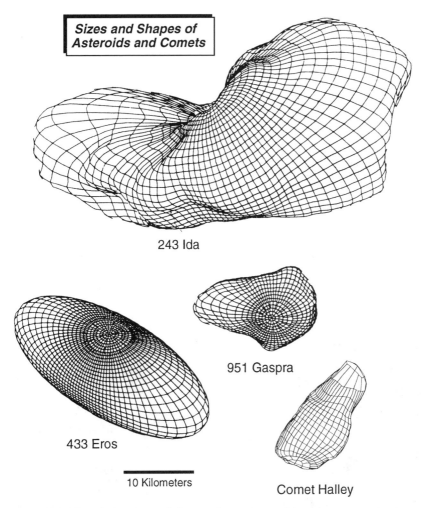

Sizes and Shapes of Asteroids and Comets

243 Ida

951 Gaspra

433 Eros

10 Kilometers

Comet Halley

Figure 3.4: The relative sizes and shapes of several asteroids and one comet nucleus have been determined from observations during spacecraft flybys. An exception is near-Earth asteroid 433 Eros, the approximate dimensions of which were determined from ground-based telescopic observations of a stellar occultation during a close approach. These images were prepared by Philip Stooke (University of Western Ontario).

that it is a **contact binary** – two separate bodies that must have bumped at very low velocity and remained in contact.

Determining the sizes and shapes of comet nuclei is more difficult than for asteroids, for the obvious reason that comet nuclei are obscured by comas. The limited data on spin rates for comets suggest that they may rotate more slowly than asteroids, perhaps allowing rotation rate to serve as a discriminator between rocky asteroids and devolatilized comets.

Close Encounters

We now have spacecraft-derived images of several asteroids that may be chondrite parent bodies. In 1991 the *Galileo* spacecraft passed within 1,600 km of asteroid 951 Gaspra. This body is irregular in shape, with an average diameter of only 12 km (see Figure 3.5). Images of Gaspra showed it to have a low density of craters, which suggests a relatively young surface age, of the order of a few hundred million years. Objects of the size of Gaspra cannot survive collisions with comparably sized bodies and are thus predicted to have short lifetimes.

Several years after the Gaspra encounter, *Galileo* passed by 243 Ida, a larger asteroid with a mean diameter of 31 km. Ida's odd shape, which has been described as resembling a croissant, is even more irregular than that of Gaspra. Its heavily cratered surface dispelled the notion that Ida could be youthful, and one estimate is that it may be several billion years old. But the most surprising find was the discovery that Ida has a satellite. This tiny (1.6 km across), egg-shaped moonlet, named Dactyl, orbits approximately 100 km from Ida's center (see Figure 3.5). Ida and Dactyl probably formed

Figure 3.5: The *Galileo* spacecraft's trajectory included flybys of two asteroids, 951 Gaspra in 1991 and 243 Ida in 1993. These images are the first close-up views of asteroids. A major surprise was the discovery of a tiny satellite, named Dactyl, orbiting Ida at a distance of approximately 100 km. Photographs courtesy of NASA.

during the same impact event, when the smaller body's ejection velocity was not high enough to allow it to escape from the gravitational grasp of its larger neighbor.

Asteroid 253 Mathilde is the largest planetesimal (its visible part measures 59 km × 47 km) so far visited by spacecraft. The *Near-Earth Asteroid Rendezvous (NEAR)* spacecraft approached Mathilde within approximately 1,200 km in 1997. The asteroid is highly angular and heavily cratered, testifying to its violent collisional history.

The two moons of Mars are no larger than modest-sized asteroids: Diemos measures 10 km × 12 km × 16 km, and Phobos measures 20 km × 23 km × 28 km. Both objects have highly unusual orbits that suggest they may have been asteroids that were gravitationally captured from the nearby main belt. The tiny moons were photographed by *Mariner 9*, and a portrait of Phobos is reproduced in Figure 3.6. Phobos has a heavily cratered surface, including one relatively large crater with a diameter that is one-fifth that of the body. The crater Stickney is associated with radial

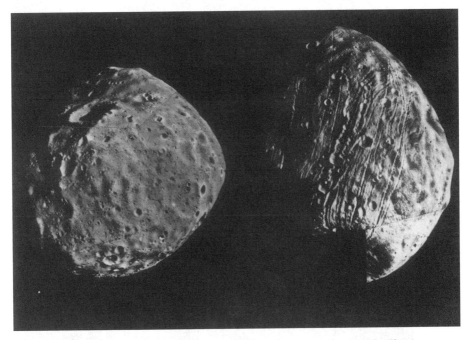

Figure 3.6: The two tiny moons of Mars are likely to be captured asteroids. These *Viking* images of Phobos from different orientations confirm the idea that violent collisions are common among asteroids. The large impact feature in shadow at the top of the right-hand photograph is crater Stickney. The fracture system on the other side of Phobos, shown on the right, suggests that the satellite was nearly disrupted by the impact that produced this crater. Photographs courtesy of Thomas Duxbury (Jet Propulsion Laboratory).

grooves that appear to be deep fractures. These features testify to a massive impact that nearly disrupted the planetesimal. The orbit of Phobos is dangerously close to Mars. Because it is still being accelerated and drawn closer, Martian tidal forces will eventually reduce it to an orbiting ring of small fragments.

Our only close-up view of a comet (see Figure 3.7) was provided by the European Space Agency's *Giotto* spacecraft, which passed within 600 km of comet Halley's nucleus in 1986. As the spacecraft raced through the hazy coma, it glimpsed a 16 km × 8 km × 8 km object, as dark as black velvet but in places violently spewing bright jets of gas and dust. Its surface is apparently covered by a deposit of carbonaceous dust, a residue left after volatilization of ices. In places this crust has cracked open, allowing ices in the interior to vaporize on exposure to sunlight. The nucleus has hills

Figure 3.7: This ghostly image of the nucleus of comet Halley was recorded by the *Giotto* spacecraft as it dashed through the coma in 1986. The nucleus is revealed to be a wobbling potato-shaped object spewing gas and dust in jets from its sunlit side. Composite image courtesy of the European Space Agency.

and depressions, and its complex, potatolike shape causes it to wobble as it rotates.

The perspective that asteroids and probably comets are parent bodies of chondritic meteorites and IDPs is in accord with our previous assessment that these materials are more or less unprocessed. By preventing asteroids from assembling into a larger planet capable of geologic processes, Jupiter may have preserved chondrites in their present, relatively pristine states. However, the irregular shapes and the highly cratered surfaces of the few asteroids that have been examined at close range indicate that such bodies have been shaped and modified by repeated collisions. The large outer planets may also have played a role in preserving cometary material by ejecting comets into the Kuiper belt before they could accrete to form larger bodies.

Another Way to Look at Asteroids

Dogs and bats can hear sounds beyond the accessible frequency range of human hearing, and many birds of prey have incredible visual capacities. We humans have learned to stretch our capacities by harnessing parts of the electromagnetic spectrum outside the range of our own limited perceptions, and a notable application is in the study of asteroids. Because asteroids generate no light of their own, this method (called **spectrophotometry**) depends on the properties of sunlight reflected from asteroidal surfaces. The ratio of light reflected by a surface to the incident light is its **albedo**, a measure of the efficiency of the reflection process. Because the measureable light also depends on the distance of the asteroid and its surface area, it is easier to obtain high-quality spectra for NEOs or asteroids in the inner main belt than for bodies of equal size at greater distances or with lower albedos.

In the discussion of isotopes and radiometric dating in Chapter 2, we considered only the protons and neutrons in the nuclei of atoms. An atom can actually be envisioned as a small pea (the nucleus) suspended at the center of a mass of cotton candy, the latter representing the distribution of electrons about the nucleus. The movement of electrons within this density cloud provides the basis for spectrophotometry. Individual electrons follow certain paths, called orbitals, but electrons can jump to outer vacant orbitals by absorbing extra energy of appropriate wavelengths. However, the energy necessary depends on the geometry of the surrounding atoms. Analysis of the energy absorbed by electrons moving between orbitals can provide information on the identity of an element and its coordination to other atoms. Sunlight provides a continuum of wavelengths from which the electrons may select. The spectrum of sunlight reflected from the surface of a mineral will be missing those wavelengths that are absorbed by electrons, the exact wavelength being characteristic for certain atoms in specific kinds of minerals. The energy ingested by crystals can have wavelengths in the visible, infrared, or ultraviolet ranges. (Light of a longer wavelength than visible is infrared; that of a shorter wavelength is ultraviolet.) For example, the most abundant element that exhibits this behavior is iron, which absorbs energy in the visible and the near-infrared parts of the spectrum when it is situated in crystallographic sites within minerals. This is the cause of visible color in many iron-bearing minerals.

Examples of the **reflectance spectra** for some individual minerals are illustrated in Figure 3.8. If these minerals are combined to form a rock,

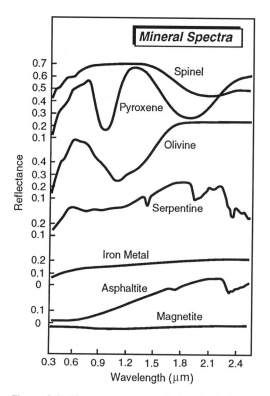

Figure 3.8: The spectrum of sunlight reflected by certain minerals shows absorption bands (valleys) at certain wavelengths because of ingestion of energy by electrons as they jump between orbitals. This forms the basis for a method of mineral identification by remote sensing. Shown here are examples of the reflectance spectra for silicates (pyroxene and olivine), hydrated silicates (serpentine), oxides (spinel and magnetite), metallic iron, and asphaltite (a complex mixture of organic compounds). Meteorites consisting of some mixture of these materials would have a composite spectrum formed by the integration of these individual spectral curves.

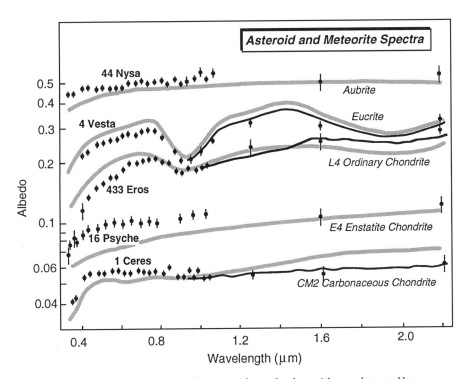

Figure 3.9: This comparison of asteroid spectra (shown by dots with error bars and by black curves) with meteorite spectra (shown by gray curves) illustrates how spectral similarities can be used to estimate the compositions of asteroids. Because absolute albedo depends on particle sizes and packing, which are unknown in the case of asteroids, it is permissible to translate asteroid spectra up or down in the diagram in order to obtain a match. Modified from published data by Clark Chapman (Southwest Research Institute).

the resulting spectrum is a messy composite of the individual spectra for the constituent minerals. These overlapping **absorption bands** provide a kind of signature for asteroidal surfaces. Examples of reflectance spectra for some asteroids are shown by dots in Figure 3.9. Each of the spectral curves has a particular shape, generally dominated by valleys characteristic of a few minerals. This is not to say that just a few minerals are present, but rather that only a few minerals produce important absorption bands. As a general rule, a linear spectrum indicates a dominant iron–nickel metal component, and a curved spectrum with absorption bands indicates silicates like pyroxene or olivine.

The spectral reflectivities of various types of powdered meteorites have also been measured in the laboratory. Spectral curves for some typical chondrites are compared with asteroid spectra in Figure 3.9. Because absolute albedo depends not only on mineralogy but also on particle sizes and packing, individual curves for meteorites and asteroids can be translated up or down in these diagrams. Consequently, the relative shapes of spectral curves are more important for comparisons than the absolute albedos. If we adjust albedo levels at some arbitrary wavelength to a common value, reasonably good matches between chondrite and asteroid spectra can be obtained in some cases.

Albedo and reflectance spectra provide the basis for a classification system that groups similar asteroids. The most widely used taxonomy was developed by David Tholen of the University of Arizona in 1984, although new classes have been recognized subsequently by other astronomers. Asteroids are assigned to one of fourteen spectral classes with letter designations (A through G, M, P through T, and V; a K class is also included in some listings). Only a subset of these groups appears to have mineralogies similar to those of chondrites, and we focus on several of them in the following sections.

S-Type Asteroids and Ordinary Chondrites

The S-type asteroids constitute the second most numerous spectral class. Asteroids 951 Gaspra and 243 Ida, both imaged by the *Galileo* spacecraft and pictured in Figure 3.5, are members of this class. The spectra of S-type asteroids provide the closest link to the ordinary chondrites, but there are some worrisome differences.

Michael Gaffey and his colleagues of Rensselaer Polytechnic Institute have carried out a systematic analysis of the spectra of the S-asteroid population. They concluded that this class is compositionally diverse and that only a small portion of all the objects lumped into this category are actually chondritic. The spectral differences among Gaffey's subtypes of S asteroids are illustrated in Figure 3.10, which plots the wavelengths of several important absorption bands. These asteroid subtypes are arrayed approximately along a curved line defined by the laboratory spectra of olivine and pyroxene mixtures in varying proportions. Only the S(IV) subclass provides a match with the spectra of ordinary chondrites; the remainder are apparently achondrite parent bodies or other asteroids for which we do not have meteorite samples.

As an illustration of the plausible identification of a specific asteroidal parent body for ordinary chondrites, let us consider 6 Hebe, a 185-km-diameter object in the main belt. Its spectrum indicates the presence of olivine, pyroxene, and iron–nickel metal, as appropriate for the mineralogy of ordinary chondrites. This asteroid exhibits minor variations in

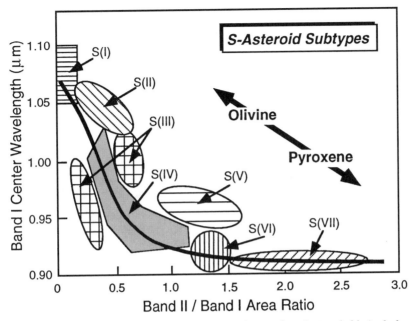

Figure 3.10: The S-type asteroids are a spectrally diverse class that probably includes both chondrite and achondrite bodies. Astronomer Michael Gaffey (Rensselaer Polytechnic Institute) has divided the S asteroids into subtypes, as illustrated in this diagram. Bands I and II refer to specific absorption bands at wavelengths of approximately 1 and 2 μm, respectively; the vertical axis is the exact position of the 1-μm band, and the horizontal axis gives the ratio of the areas occupied by both absorption bands. All these subtypes are distributed along a curve representing mixtures of olivine (plotting at the upper left) and pyroxene (plotting at the lower right). Only the S(IV) class appears to be spectrally similar to ordinary chondrites.

its spectra as it rotates, suggesting some mineralogic differences on its surface. Spectra taken at different times during Hebe's rotation are illustrated in Figure 3.11, but all the spectra plot within the S(IV) region of this diagram. Originally it was argued that rotational spectral variations indicate that an asteroid cannot be chondritic because chondrites by definition were thought to be compositionally uniform. However, the observed variations can be explained by small changes in the relative proportions of olivine and pyroxene resulting from chondrite metamorphism. The clustering of Hebe's spectra at the lower end of the S(IV) field matches the spectra of H chondrites, and the observed rotational variations are similar to those measured for different metamorphic grades of H chondrites.

Although the positions of the absorption bands for Hebe and H chondrites correlate quite well, the albedos of these objects do not, as shown

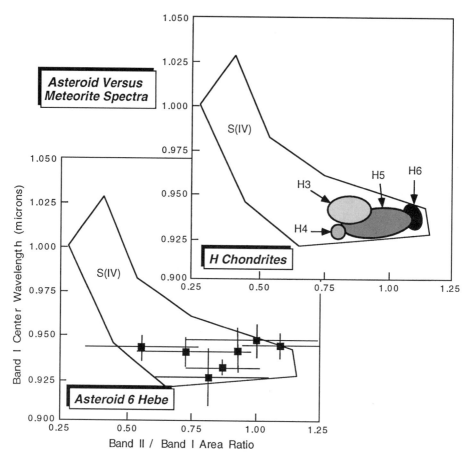

Figure 3.11: These expanded views of the S(IV) field from Figure 3.10 show a comparison of the spectra of asteroid 6 Hebe (squares with error bars) and H chondrites of various petrologic types (shaded ovals). Other classes of ordinary chondrites plot in other parts of the S(IV) field. Variations in the Hebe spectra result from the asteroid's rotation, which presents regions with different spectra at different times. The observed variation matches approximately the trend resulting from changes in the olivine-to-pyroxene ratio caused by metamorphism. The spectral similarities suggest that 6 Hebe may be the H-chondrite parent body. Modified from the published work of Michael Gaffey (Rensselaer Polytechnic Institute) and Sarah Gilbert (National Institute of Standards and Technology).

in Figure 3.12. The Hebe spectrum at wavelengths longer than approximately $0.6\,\mu\text{m}$ is less curvaceous and more highly reflective (reddened, in the parlance of astronomers). A number of possible explanations have been offered to explain this reddish blemish, but the most likely seems to be **space weathering**. When an asteroidal surface is exposed to the interplanetary space environment for millions of years, it undergoes chemical and physical changes as a result of bombardment by meteoroids and

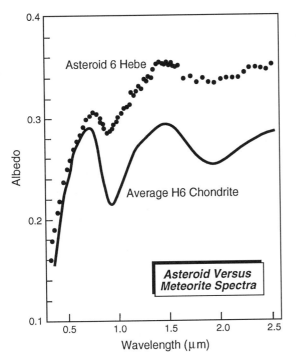

Figure 3.12: Despite the impressive spectral match between asteroid 6 Hebe and H chondrites shown in Figure 3.11, some differences remain. In particular, the Hebe spectrum is more reflective at higher wavelengths (reddened, in the vernacular of astronomers) than are the spectra of H chondrites. This difference probably reflects the effects of space weathering, an alteration process that is thought to modify the spectral characteristics of surface layers on asteroids.

interaction with the solar wind. These processes have clearly produced spectral reddening of the Moon's surface, and space weathering of asteroidal surfaces might account for a similar change in their spectra. *Galileo* high-resolution images of 243 Ida appear to provide further proof of space weathering on asteroids. Ida's ancient surface is reddened, except in the vicinity of younger craters that have excavated materials that have not been exposed on the surface for very long. Yet another observation is equally suggestive. Richard Binzel and his colleagues at the Massachusetts Institute of Technology have discovered a handful of S-type objects in the near-Earth asteroid population, and their spectra fill the gap between the spectra of Hebe and H chondrites. Some of these NEOs may be fragments of Hebe, dislodged at various times so that their surfaces have experienced varying degrees of space weathering.

Specific parent bodies for the other groups of ordinary chondrites have not yet been identified. One small (7-km-diameter) main-belt asteroid,

3628 Boznemcová, exhibits a spectrum similar to that of L and LL chondrites, but it seems too small to be more than a mere fragment of the original parent body for one of these chondrite classes.

C-Type Asteroids and Carbonaceous Chondrites

Dim asteroids are very abundant in the asteroid belt, but their low albedos present a special problem for spectrophotometry. The C-type asteroid 253 Mathilde offers a good example. Its portrait, constructed from mosaic images obtained by the *NEAR* spacecraft, reveals an object nearly twice as dark as a charcoal briquette (see Figure 3.13). Mathilde reflects only approximately 5% of the incident sunlight, so obtaining its spectrum is challenging. The slopes and the floors of large craters on Mathilde exhibit the same black coloration, suggesting that the body may be homogeneous throughout.

The most intensely studied such objects are the two largest asteroids, 1 Ceres and 2 Pallas. The C-type asteroids (and their relatives, the G, B, and F types) have nearly featureless spectra at wavelengths longer than $0.4\,\mu$m, but their infrared spectra (near $3\,\mu$m) are revealing. Most of these

Figure 3.13: This global portrait of C-type asteroid 253 Mathilde, constructed from four images taken by the *NEAR* spacecraft in 1997, shows a body that is uniformly blacker than charcoal. Such low albedos make C-type asteroids challenging targets for spectrophotometry. Photograph courtesy of the Applied Physics Laboratory of Johns Hopkins University and NASA.

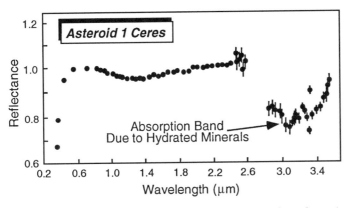

Figure 3.14: The spectrum of asteroid 1 Ceres shows a deep absorption band in the infrared region at a 3-μm wavelength. This feature is attributed to the presence of water in hydrated minerals like clays. The gap in the spectrum between 2.6 and 2.8 μm is due to absorption of the signal by water in the Earth's atmosphere. These measurements were obtained by Thomas Jones and colleagues (University of Arizona).

dark asteroids show a prominent absorption band at this position (see Figure 3.14), indicating the presence of hydrated minerals such as phyllosilicates. Water bound into the structures of clay minerals vibrates when it absorbs energy of this wavelength, but this feature in asteroid spectra is not always easy to detect because of interference as the signal passes through the Earth's atmosphere. Some C-type asteroids also exhibit other weak absorptions caused by iron in phyllosilicates and possibly by organic matter. All these spectral features and the compositions they suggest are similar to those of carbonaceous chondrites, especially the CI and CM chondrites that have suffered aqueous alteration.

The amount of water in phyllosilicates, as judged from the depth of the 3-μm absorption band, appears to decrease in C asteroids with increasing size. This may suggest that some of the larger dark asteroids may represent the heated cores of once-larger parent bodies and that most CI and CM chondrites may have been derived from the lost outer portions. 1 Ceres (which is actually a member of the G class) is a notable exception, since its surface shows a prominent water absorption band. The spectrum of 2 Pallas closely resembles that of CR chondrites, possibly suggesting that it may be the parent body for this chondrite group.

D- and P-Type Asteroids, Centaurs, and Interplanetary Dust

D and P asteroids are among the darkest objects known. Their spectra are generally featureless with very reddened reflectance at long wavelengths. Because of their low albedos and reddened spectra, these asteroids are generally believed to contain large quantities of organic matter and opaque minerals like magnetite. However, they do not generally show the 3-μm absorption feature attributable to hydrous alteration minerals, so any water they contain must still be in frozen form.

5145 Pholus, lying just outside the orbit of Saturn, is one of the reddest objects in the solar system with a spectrum unlike that of any known asteroid. The best spectral match appears to be provided by mixtures of complex organic matter and ices. Its surface could be a crust of organics and silicate dust, perhaps something like that of comet Halley. Pholus is probably a wandering Kuiper belt object, similar in many respects to comets.

No known meteorites have spectra or mineralogies that resemble D- or P-type asteroids or Centaur objects. In 1996, mineralogist John Bradley and his collaborators published the first reflectance spectra for interplanetary dust particles. Acquiring these data was something of a technological feat, because the particles are so tiny. IDPs that are dominated by phyllosilicates not surprisingly exhibit spectra that resemble C-type asteroids, and these particles were probably derived from such bodies. More interesting, though, are the CP IDPs that contain no hydrated minerals. These particles are rich in carbonaceous material and show very reddened spectra similar to those of D and P asteroids and Pholus.

These results provide support for the idea that the micrometeorite collection contains samples from parent bodies that are not represented in collections of larger meteorites. It also seems likely that at least some CP IDPs are cometary in origin. Comet dust particles have orbits that require higher velocities than those for particles derived from asteroids. IDPs entering the atmosphere at higher speeds are heated to higher temperatures and thus are purged of the gases they contain more efficiently. Physicist Al Nier of the University of Minnesota measured the temperatures at which helium was released from IDPs, thereby identifying the severely heated motes that are more likely to be cometary dust.

A Thermal Gradient and a Snow Line

Asteroid spectra in general are more varied than the spectra of known meteorite types. This reinforces the idea that meteorites have so far provided only a limited and probably biased sample of what is out there. Moreover, the relative abundances of asteroidal types are nothing like those of fallen chondrites. C-type asteroids far outnumber all other types, possibly composing three-fourths of the main belt, in contrast to the relative scarcity of carbonaceous chondrites. S-type asteroids, especially members of the

S(IV) subclass that are probable parent bodies for the abundant ordinary chondrites, make up only a small fraction of the belt. And D- and P-type asteroids are apparently not sampled at all, except as minute dust particles. We cannot assume that a relatively unimportant group of meteorites or IDPs is not volumetrically important in the asteroidal source region.

The asteroid classes thought to be chondritic in composition are not distributed uniformly within the main belt, as illustrated in Figure 3.15. The inner belt inside 2.5 AU is dominated by the diverse S-type asteroid class. Asteroid 6 Hebe, which may be a likely candidate for the H-chondrite parent body, has a mean orbital distance of 2.4 AU. The middle belt is dominated by C-type asteroids and their dark relatives, so this is the likely domain for carbonaceous chondrite parent bodies. Out past 3.5 AU are the P- and the D-type asteroids (most Trojan asteroids, located at 5.2 AU, are D type), and farther still are the highly reddened Centaurs, Kuiper belt objects, and comets. These distant bodies are sampled, if at all, only as microscopic specks of interplanetary dust.

An interesting inference that can be drawn from this regular pattern is that most objects inside approximately 2.5 AU contain little or no water, whereas those at greater heliocentric distances contain either hydrous minerals or ices. This has led to the concept of a **snow line**, which defines the solar distance at which volatile ices condensed in the solar

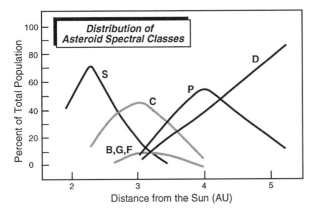

Figure 3.15: Synthesis of available reflectance spectra suggests that the proportions of different asteroid classes vary systematically with heliocentric distance. S-type asteroids are concentrated in the inner main belt, and C-type asteroids and their dark relatives (the B, G, and F types) occur in the middle of the belt. These S- and C-type objects may be the sources of ordinary and carbonaceous chondrites, respectively. The very dark P and D asteroids occur primarily in the outer belt and may be sampled only as IDPs.

nebula. We might more accurately rename asteroids outside the snow line as iceteroids.

There is also an apparent gradation in the thermal histories of asteroidal bodies. In the inner main belt, most objects were heated rather severely, causing metamorphism or, as we see later in Chapter 4, melting. Temperatures in the middle of the main belt were sufficient to cause melting of the accreted ice, which reacted with silicates to form hydrated phyllosilicates and other alteration minerals. At distances greater than approximately 4 AU, temperatures remained low enough that even ices were apparently preserved intact.

Asteroidal parent bodies for chondrite types that we have not discussed, such as the enstatite or the Rumuruti chondrites, have not yet been positively identified. From the meteorite properties, some scientists have inferred that they formed inboard of the ordinary chondrite parent bodies. Unfortunately, most of the minerals comprising these chondrite classes do not have distinctive spectral signatures.

Warming an Onion

Fragments in chondritic breccias, although usually of the same meteorite group, commonly differ in terms of metamorphic grade (petrologic type). This observation can be interpreted in two ways: Either colliding planetesimals were usually of the same group but had been heated to different degrees or similar impacting bodies contained both metamorphosed and unmetamorphosed materials. The second idea seems much more plausible. Rocks are notoriously poor conductors of heat, and as a consequence heat was unlikely to have been distributed evenly throughout the interiors of even small asteroids. This would have produced zones of differing metamorphic grades arranged radially within each body.

Postulated heat sources for metamorphism were the rapid decay of short-lived radioactive isotopes, impacts, or electric currents induced by the early solar wind. The decay product of a short-lived and now extinct isotope of aluminum, ^{26}Al, has been found in refractory inclusions in chondrites. This isotope was apparently uniformly distributed with high enough abundance in the nebula to cause metamorphism of accreted bodies, but the timing of accretion was critical (too quickly caused the bodies to melt and too late stifled metamorphism). Although impact has been advocated as a heat source, this mechanism is effective only for bodies larger than asteroids. Resistance to the flow of electric currents induced by a ferocious solar wind could conceivably cause metamorphism, but this idea depends critically on a model of an early T Tauri sun that is no longer widely supported. Moreover, the process may be self-limiting, because resistance to current flow decreases as temperature rises. The only heating mechanism for which chondrites provide any observational support is the decay of short-lived radionuclides, and most thermal models are based on this premise.

Before we examine some thermal models for chondrite parent bodies, let us briefly review some facts we have learned about metamorphism from studies of ordinary chondrites. Peak temperatures in chondrite parent bodies ranged from approximately 750–950 °C for heavily metamorphosed type 6 chondrites to 400–600 °C for relatively unmetamorphosed type 3 chondrites. If the relative abundances of chondrite types have any significance, more highly metamorphosed chondrites predominate over unmetamorphosed ones. Chondrite age data, based on uranium–lead isotopes (which provide higher resolution than the rubidium–strontium chronometer) indicate that the time interval between parent body formation and the end of metamorphism was only approximately 60 million years. If we accept this interval as the time required for metamorphosed chondrites to cool to the approximate blocking temperature for the movement of lead atoms in minerals, we can estimate a cooling rate of a few tens of degrees per million years. This value can be combined with measurements of the rate at which heat flows through chondrites to estimate the size of ordinary chondrite parent bodies. Typical values for the diameters of such bodies calculated in this way are several hundred kilometers, very reasonable sizes for asteroids.

Detailed thermal models for ordinary chondrite parent bodies have been formulated with the foregoing constraints and the assumption of internal heating by ^{26}Al decay. The most tightly constrained model is for the H-chondrite parent body. The asteroid accreted several million years after formation of refractory inclusions, which are thought to be the earliest-formed objects in the solar nebula. The thermal model predicts an asteroid diameter of 175 km (which is, incidentally, in good agreement with the 185-km size of 6 Hebe, previously discussed as a possible H-chondrite parent body). After accretion, its internal temperature began to rise, as the heat from decay of ^{26}Al was generated faster than it could be conducted to the surface and escape to space. Peak temperature in the center of the body was reached approximately 3 million years after its accretion, after which time it began to cool slowly. A schematic model of the H-chondrite parent body from such calculations is shown in the left-hand side of Figure 3.16. Thermal models of parent bodies like this with concentric zones of different metamorphic grades are called **onion shell models**. From the relative proportions of the various petrologic types in such

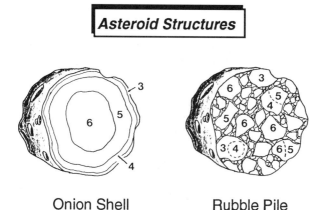

Asteroid Structures

Onion Shell   Rubble Pile

Figure 3.16: These sketches compare two models for the internal structures of chondrite parent bodies. The onion shell body is thought to result from heating by rapid decay of short-lived radioactive isotopes. This internal heating produces a heavily metamorphosed (type 6) interior, but slow conduction of heat allows the near-surface layers to remain relatively cool. The rubble-pile structure can be produced when onion shell bodies collide catastrophically. After disruption, the fragments are reassembled randomly by mutual gravitational attraction into a new body. In this way, materials at different metamorphic grades may be exposed on the asteroid's surface.

calculations, we can readily see why heavily metamorphosed chondrites are so common.

If metamorphism occurred within internally heated bodies, some highly volatile elements mobilized by heat should probably have migrated to cooler regions near the surface, where they might have been redeposited. Very high contents (sometimes above cosmic abundances) of easily mobilized volatile elements such as bismuth, indium, and thalium have been measured in some type 3 ordinary chondrites that presumably were situated near the surfaces of their parent bodies. In this way, even unmetamorphosed chondrites may have felt some effects of thermal metamorphism happening deep in the interiors of their parent asteroids.

If the rate of accretion had been faster, metamorphic temperatures could have been reached earlier in smaller bodies with diameters of just a few kilometers. Electromagnetic induction heating would also work more efficiently on smaller objects. Are there other means by which we can constrain the sizes of metamorphosed chondrite parent bodies? Pressure increases with depth in a planetesimal because of the mass of the overlying rock. Pressure at any point within the body is a function of rock density and gravitational acceleration, in turn related to the size of the body. For chondrite parent bodies with diameters of a few hundred kilometers, the most extreme pressure experienced at its center is less than

0.15 GPa, approximately 1,500 times the Earth's surface atmospheric pressure). There are some minerals or mineral assemblages whose compositions are pressure sensitive, but unfortunately these are applicable only at higher pressures. Substitution of extra sodium and aluminum in pyroxene is dependent on pressure, and small increases in these elements occur in chondritic pyroxenes with increasing metamorphic grade. The total amount of this component suggests pressures of less than 0.1 GPa. Such data are consistent with an internally heated body, but the low pressures occurring in asteroid-sized objects have so far precluded an accurate assessment of their sizes during metamorphism.

Postmetamorphic cooling rates are also related to size and position within the body. In principle, material deep in the interior of a body is effectively insulated and should cool more slowly than that at or near the surface. This effect is apparently seen in the measured uranium–lead isotopic ages of ordinary chondrites. Ages for ordinary chondrites measured by isotope chemist Christa Göpel and her colleagues range from 4.56 to 4.50 billion years, with the youngest ages for heavily metamorphosed meteorites. If situated in the deep interior of an onion shell body, these rocks would cool more slowly and thus reach the blocking temperature at a later time.

There are also several ways in which cooling rates can be assessed directly; in effect, these are **cooling-rate speedometers**. One method for determining cooling rates uses the fission tracks that were noted in Chapter 2 as evidence for decay of some extinct radionuclides. Although these tracks are produced continuously, minerals at high temperatures anneal themselves and erase these tiny imperfections. Different minerals, however, begin to retain tracks at different temperatures. For example, the track retention temperature for pyroxene is approximately $300\,^{\circ}C$, and that for whitlockite (calcium phosphate) is several hundred degrees lower. If the tracks were produced by the decay of ^{244}Pu, then from the density of tracks and the known decay rate we can estimate the time interval for the meteorite to have cooled from 300 to $100\,^{\circ}C$. A summary of cooling-rate curves based on ages from fission track retention and blocking temperatures of radiogenic isotopes in H chondrites is shown in Figure 3.17. The most severely metamorphosed chondrites exhibit the slowest cooling rates, supporting the idea that they were the most deeply buried. These different rates reinforce the hypothesis that the metamorphism of ordinary chondrites occurred in fairly large parent bodies with onion shell structures.

Thermal models for carbonaceous chondrite parent bodies are not as well constrained as for ordinary chondrite asteroids, but they provide

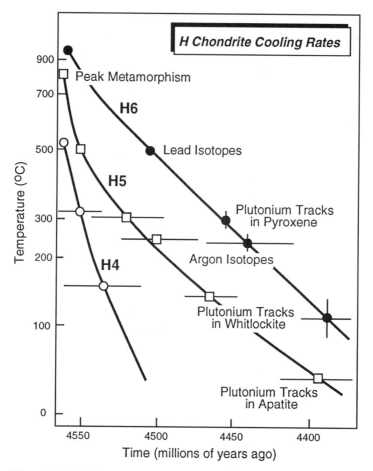

Figure 3.17: Cooling rates determined for ordinary chondrites provide support for the existence of onion shell bodies. These data, summarized from the work of Paul Pellas (Paris Mineralogy Museum), show that strongly metamorphosed chondrites (Guarena, Spain, H6) cooled more slowly than less metamorphosed chondrites (Richardton, North Dakota, H5 and Ste. Marguerite, France, H4). The slopes of the curves are defined by the times that various isotopic systems were frozen, as well as times determined by retention of plutonium fission tracks in various minerals.

important insights into the evolution of ice-bearing planetesimals. Aqueous alteration generally occurred at fairly low temperatures, probably less than 100 °C. Models indicate that ice acted as a buffer – as long as frozen water was present, its melting absorbed most of the heat generated by ^{26}Al decay, moderating the temperature rise in asteroids. Once the ice supply was exhausted, however, temperatures climbed, perhaps accounting for the existence of some altered carbonaceous chondrites with clays that were subsequently heated and dehydrated.

Bump and Grind

Many asteroids have highly irregular shapes, as detected from their tumbling motions inferred from observed rotational changes in brightness. We have already seen that all the asteroids visited by spacecraft have angular shapes, suggesting that they are collisional fragments. The sizes of asteroids, too, show considerable scatter. Astronomers have noticed a curious distinction between the size distributions for C- and S-type asteroids. When plotted on a logarithmic diagram of diameter versus abundance, C-type asteroids form a straight line. Such a linear relation is known to be characteristic of a population of objects whose sizes have been determined by fragmentation that is due to mutual collisions. When S-type asteroids are plotted on the same diagram, they exhibit a nonlinear trend, with a hump, diagnostic of incomplete fragmentation. We therefore expect that C-type asteroids have been more thoroughly disrupted by repeated impacts. This is a logical outcome of the fact that carbonaceous chondrites have lower crushing strengths than ordinary chondrites. A few chondritic bodies, such as 1 Ceres, so far appear to have escaped the collisional fate of most other asteroids.

Ordinary chondrite parent bodies provide an especially instructive lesson in the effects of asteroid collision. The H-chondrite parent asteroid appears to have been destroyed by a massive impact and then gravitationally reassembled from the resulting fragments. This conclusion comes from study of a cooling-rate speedometer based on metal grains. **Metallographic cooling rates** depend on the diffusion of nickel from one iron–nickel metal alloy (taenite) into the other (kamacite) at high temperatures. Nickel moving to the edges of taenite grains forms a compositional gradient that can be measured. Movement of nickel atoms ceases below approximately 500 °C, so the nickel compositional gradient is frozen in at this temperature. From measured profiles of nickel concentrations in taenite, it is possible to calculate the rate at which these grains cooled. Cooling-rate contours are illustrated in Figure 3.18. The Cangas de Onis (Spain) meteorite is a breccia composed of H-chondrite fragments of differing metamorphic grades. Metallographic cooling rates determined from nickel profiles in individual taenite grains in this breccia indicate widely varying cooling rates, as we can judge by comparing their compositions with the cooling-rate contours in Figure 3.18. The rates, ranging from as fast as 1,000 °C to as slow as 1 °C per million years, correspond to burial depths spanning the interval from the surface to 100 km. Based on evidence we have already discussed for the size of the H-chondrite parent body, this one meteorite breccia appears to contain

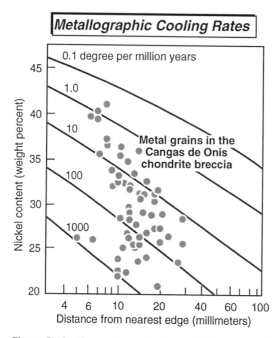

Figure 3.18: The cooling rates for individual grains of iron–nickel metal in the Cangas de Onis (France) H-chondrite breccia vary from 1000 to 1 °C per million years, corresponding to burial depths from near the surface to 100 km inside the parent body. The inclusion of rock fragments from such depths in this breccia implies that the original asteroid was catastrophically disrupted and reaccreted into a rubble pile. These data were collected by Jeffrey Taylor and colleagues (University of Hawaii).

rock fragments extracted from the near surface and from its deepest interior. It is inconceivable that an asteroid could survive an impact that sampled its center, so it must have been disrupted and reassembled into a new object, now perhaps recognizable as asteroid 6 Hebe. The interior of such a body would comprise a jumbled assortment of rocks of various metamorphic grades taken from throughout the original onion shell body, a structure that has been termed a **rubble-pile model** (see Figure 3.16). This structure might explain the observed rotational variations in Hebe's spectra, because it allows highly metamorphosed blocks to be randomly exposed at various places on the surface.

The L-chondrite parent body was apparently also catastrophically disrupted by impact but, unlike the H-chondrite asteroid, it probably did not reaccrete. As a class, L-chondrites are more severely shock metamorphosed than other chondrite types, and the proportion of shocked and partly melted L-chondrites (88%) is much higher than would be expected from a typical cratering event. The timing of parent body disruption can be ascertained from chondrite gas-retention ages, which stipulate the time when a gaseous isotope of argon, ^{40}Ar, began to accumulate from the decay of radioactive potassium, ^{40}K. All heavily shocked L-chondrites have gas-retention ages of approximately 500 million years, marking the time of this event. Asteroid 3628 Boznemcová, which has a spectral signature similar to that of L6 chondrites, may be a relatively small fragment surviving from this collision.

Determination of the orbit of the satellite Dactyl around asteroid 243 Ida allows the mass of Ida to be estimated. When divided by its volume, this gives a density for the body of 2.6 ± 0.2 g/cm^3. This value is slightly less than that for coherent chondrites (with typical values of approximately 3 g/cm^3), suggesting that Ida, too, may be a rubble pile. The argument is much stronger for 253 Mathilde with a density, estimated from its volume and gravitation influence on the *NEAR* spacecraft, of only 1.3 ± 0.2 g/cm^3, not much more than the density of water. Although incorporation of a large amount of ice could explain this low densitiy, Earth-based spectroscopic measurements of the asteroid show no signature of ice. Mathilde must have a porosity of approximately 50%! Furthermore, the rates at which many other asteroids spin seem to provide support for the idea that most may be collections of orbiting rubble. Below a critical density, a rapidly rotating body will fly apart; only coherent bodies can withstand the tensile forces generated by rotation. The spin rates for small asteroids are all very modest, and a plot of these rates appears to be truncated at the rotational velocity at which rubble piles would disintegrate.

Chondritic Dirt

Scientists once visualized the lunar surface as marred by sharp fractures and angular, rocky prominences. The first high-resolution photographs form the Moon, however, indicated a subdued, rolling topography. The reason for this miscalculation is a mantling layer of soil, called the **regolith** (Greek for rock layer). Regolith is defined as a pervasive blanket of loose, rocky material that rests on coherent bedrock. The *Apollo* astronauts noted that their rocket exhausts created flurries of rock dust, often as fine as flour (see Figure 3.19). However, soil samples collected on the Moon also contain blocks of various sizes. Regoliths form because of repeated bombardment by small meteoroids that pulverize the

Figure 3.19: These footprints on the lunar surface were made by *Apollo 15* astronauts. The sharpness of the indentations shows that the soil consists of fine, cohesive powder. The regolith also contains blocks of rock and was formed during continued impacts by meteoroids of all sizes. A similar regolith is probably characteristic of chondritic asteroids. Photograph courtesy of NASA.

surface rocks, occasionally punctuated by larger impacts that excavate blocks from below and mix them with dust.

This soil-forming process is not restricted to the Moon. *Viking* and *Mars Pathfinder* lander photographs of the Martian surface indicate development of a regolith, and such features are probably characteristic of all planets with cratered surfaces. The asteroid-sized satellites of Mars have regoliths (see Figure 3.20), and it seems reasonable to assume that most asteroids do as well. Polarization of light reflected from asteroids has been interpreted as indicating dust-covered surfaces of unknown thickness. *Galileo's* images of 951 Gaspra suggest only a thin regolith, but the regolith on 243 Ida has been estimated at 150 m deep, approximately the same thickness if all ejecta from its large craters were retained and spread evenly about the surface.

At present, our best source of information on asteroidal regoliths comes from chondrites. Even if loosely consolidated regolith samples could be lauched into space and reach the Earth, they could not survive atmospheric transit (except, of course, as individual particles of interplanetary dust). Fortunately, well-indurated clumps, called **regolith breccias**, sometimes form because of compaction and cementation of the various soil ingredients. Such regolith breccias comprise approximately half of the carbonaceous chondrites and smaller percentages (generally 10% or

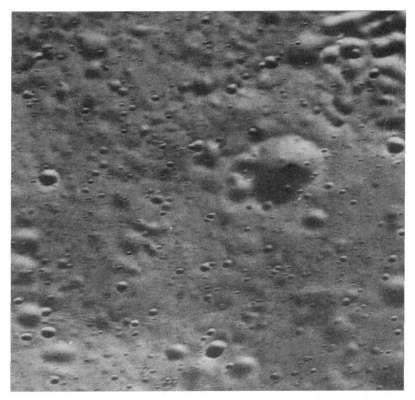

Figure 3.20: Although they have weak gravitational fields, small solar system bodies can apparently retain enough comminuted impact ejecta to form thick regoliths. That this is possible is demonstrated by a close-up view of the surface of Phobos, one of the tiny moons of Mars photographed by the *Viking* spacecraft. A regolith blankets the landscape and produces a subdued appearance despite the presence of craters. Photograph courtesty of Thomas Duxbury (NASA's Jet Propulsion Laboratory).

less) of other chondrite classes. However, not all chondritic breccias were formed in regoliths. Remember that many asteroids may be rubble piles, in effect, giant breccias reassembled from smashed asteroids, and most of these materials would be buried below the zone of active regolith formation. Only the outer veneer of breccias would be reworked into regolith.

The textures of all breccias, regolith and otherwise, are similar and consist of angular rock and mineral fragments of varying sizes. However, true regolith breccias display some distinctive characteristics that allow us to recognize their mode of formation. On bodies without atmospheres, surficial material is subjected to continuous bombardment by energetic nuclear particles. This **cosmic irradiation** has left unmistakable fingerprints in some chondritic breccias. Significant amounts of gases with the distinctive isotopic composition of the solar wind were implanted on the surfaces of individual grains. Microscopic craters and particle tracks also formed as more energetic ions penetrated into grains. An array of new isotopes was further produced by the disruption of incoming particles, and these can be measured. Because cosmic rays and solar wind provide particles with different energies, the irradiation effects of the more energetic particles extended to greater depths in regoliths. The record of exposure to cosmic particles with different energies can even be converted into an **exposure age** for each type of radiation. Typical times for surface exposure of chondrite regolith materials are generally 10 to 50 million years.

One interesting feature of regoliths is that their characteristics change, or mature, as their surface residence times increase. Exposure ages for lunar regolith samples are generally an order of magnitude longer than those for chondrite breccias. The greater age of lunar regolith breccias affords an instructive comparison with chondritic regolith samples. The most visible difference in maturity between these is the presence in the lunar regolith of glass and partially melted rock and soil fragments, called **agglutinates**. These melted products are the result of continued impacts, which increase with exposure time. Glass and agglutinates make up as much as 60% of mature lunar breccias, but they are rare in chondrites. The lunar regolith is also much finer grained than its chondritic counterparts.

Why should the surface residence time and maturity of regolith materials be so much less on asteroids than on the Moon? One possible explanation hinges on the fact that the amount of orbiting material and the resulting impact rates are much greater in the asteroid belt than in the vicinity of the Moon at 1 AU; consequently, more impacts might result in higher rates of regolith stirring and overturn. Asteroids have weaker gravity fields than the Moon because they are less massive, and their short regolith exposure ages have also been attributed to easier ejection of regolith materials into space. Mathematical models confirm that impact rate and parent body size are the controlling factors in regolith formation, but point to a different and rather surprising conclusion. These calculations suggest that asteroidal regoliths are much thicker (of the order of 1 km for 100–300-km-diameter bodies) than the lunar regolith (measured at approximately 5–10 m at the *Apollo* landing sites). The greater thickness results in part from the fact that the higher lunar gravity produces less widespread ejecta. The same impact onto an asteroid-sized body would deposit ejecta over the entire object because of its lower gravity field.

Thus the explanation for the shorter exposure time for chondrite regolith breccias may be that surface materials are buried faster under more rapidly evolving surfaces.

Processes Affecting Comet Nuclei

A common perception of comets is that they are agglomerations of primordial dust and ices that have been perfectly preserved in the cold storage of distant orbits. However, short-period comets, which are the ones we are most likely to examine and sample, have probably been modified by some of the same processes that have affected asteroids, although probably to a lesser degree (see Figure 3.21). Even comets in the Oort cloud can be warmed by supernovae and irradiated by galactic cosmic rays.

When comets enter the planetary region, increased solar warming can affect the original amorphous ices. Condensed ices are thought to lack regular crystal structures, but these can readily transform into crystalline ices with only mild heating. Spectral observations suggest that comet Halley may have reached temperatures as high as 50 °C as it passed inside the Earth's orbit, which is several hundred degrees warmer than the temperature required for crystallizing amorphous ices on or near the sur-

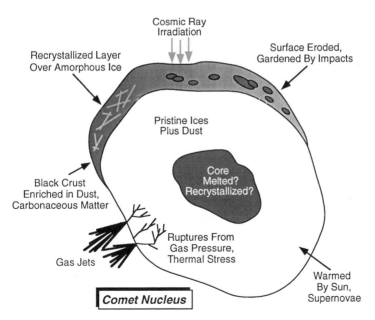

Figure 3.21: As illustrated in this figure, a comet nucleus may be subjected to a variety of processes that can alter its surface and interior.

face. If comets were also heated by decay of short-lived radionuclides in accreted dust, their cores could be melted and recrystallized as well. Heating by whatever mechanism would produce gases that could rupture the outer layers of the nucleus and be vented into space. Continued loss of volatiles from surface layers would result in a coating enriched in silicate, oxide dust, and organic matter, probably similar to the black crust observed on comet Halley.

Although the spacings between comets in the Oort cloud or Kuiper belt are great enough to minimize impacts, the existence of smaller nuclei in these regions might allow collisions to occur. Asteroid surfaces could thus be eroded by impacts, and hardened surface deposits might bear some similarities to asteroidal regoliths. Irradiation by cosmic rays and the solar wind presumably produce particle tracks and nuclear reactions like those in asteroid targets. Laboratory experiments simulating irradiation of dust and ice mixtures suggest that carbon compounds may combine to form complex organic polymers to depths of a few meters in cometary nuclei. These organic compounds may function as a sticky glue that seals the nucleus against volatile loss until it is ruptured by thermal stresses and gas pressure.

Junk or Treasure?

Archeologists must reconstruct ancient settlements and infer something about the past lifestyles of lost civilizations, often from painfully little evidence. Many of their best clues are provided by what once was worthless garbage cast aside by the populations under scrutiny. A few chips of pottery, a broken kitchen utensil or arrowhead, or some bones or seeds can provide valuable information on trade, industry, and diet.

In an analogous manner, we are forced to reconstruct or infer the salient features of the parent bodies for chondrites and interplanetary dust from limited astronomical and spacecraft observations and the accumulated record of postaccretional processes in the samples themselves. Neither of these methods of inquiry alone provides a comprehensive view of where these materials originate, so an interdisciplinary approach is necessary. We have learned from orbital measurements that asteroids are probably the sources of chondrites, and the spectral properties of some individual asteroids match those of specific kinds of meteorites. The parent bodies of distinct chondrite groups apparently formed under different nebular conditions prevailing at various distances from the Sun. Thermal processes resulted in metamorphism or aqueous alteration of planetesimal interiors, and collisions between asteroids altered their sizes and shapes and sometimes produced reaccreted piles of rubble. Impacts on smaller scales, as well as irradiation by cosmic rays and the solar wind, created regoliths blanketing their surfaces. These space-weathering processes sometimes obscure spectral characteristics. Some IDPs are derived from objects not sampled as meteorites, probably dark, distant asteroid types and comet nuclei.

We began this chapter by noting that asteroids and comets are essentially debris left over from the formation of the planets. The insignificant sizes of these small bodies belie their importance in solar system history. Planetesimals of chondritic composition provide our earliest historical records, and their constituents reflect the most fundamental processes and events that we can hope to study.

Suggested Readings

There are relatively few nontechnical publications on asteroids, but many on comets. Many of these papers are demanding but excellent sources.

GENERAL

Balsiger H., Fechtig H., and Geiss J. (1988). A close look at Halley's comet. Sci. Am. **259**, 96–103.

This easily readable article summarizes the results of Giotto's mission to the nucleus of comet Halley.

Beatty J. K. and Chaikin A., eds. (1990). *The New Solar System*, 3rd ed., Sky Publishing, Cambridge, MA.

This colorful book contains pertinent chapters on asteroids by C. R. Chapman and W. K. Hartmann and on comets by R. Reinhard and J. C. Brandt.

Binzel R. P., Barucci M. A., and Fulchignoni M. (1991). The origins of the asteroids. Sci. Am. **265**, 88–94.

An excellent and easily understood article describing the spectral classification and physical properties of asteroids.

Sagan C. and Druyan A. (1985). *Comet*, Random House, New York.

This is a highly entertaining and beautifully illustrated book describing what was known about comets before the Giotto encounter with Halley.

ASTEROID AND COMET PROPERTIES AND ORBITS

Binzel R. P., Gehrels T., and Matthews M. S., eds. (1989). *Asteroids II*, University of Arizona, Tucson, AZ.

This massive tome, nearly three inches thick, summarizes most of what is known about asteroids. Part II, containing sections on exploration, and Part

III, on structure and physical properties, are especially relevant to this chapter's topics.

Newburn R. L., Neugebauer M., and Rahe J. (1989). *Comets in the Post-Halley Era*, Kluwer, Dordrecht, The Netherlands.

This is a two-volume collection of scientific papers following an International Astronomical Union colloquium on comet Halley.

SPACECRAFT ENCOUNTERS WITH ASTEROIDS AND COMETS

Special issue: Comet Halley (1986). Nature (London) **321**, 259–366.

A collection of technical papers describing Giotto's encounter with the nucleus of comet Halley.

Sullivan R., Greeley R., Pappalardo R., Asphaug E., Moore J. M., Morrison D., Belton M. J. S., Carr M., Chapman C. R., Geissler P., Greenberg R., Granahan J., Head J. W., Kirk R., McEwen A., Lee P., Thomas P. C., and Veverka J. (1996). Geology of 243 Ida. Icarus **120**, 119–139.

An assessment of findings from Galileo's flyby of Ida; reports dealing with spacecraft missions tend to have very long author lists!

Veverka J., Belton M. J. S., Klaasen K., and Chapman C. R. (1994). *Galileo's* encounter with 951 Gaspra: overview. Icarus **107**, 2–17.

A summary of results from the first spacecraft flyby of an asteroid.

ASTEROID SPECTROPHOTOMETRY AND TAXONOMY

Bradley J. P., Keller L. P., Brownlee D. E., and Thomas K. L. (1996). Reflectance spectroscopy of interplanetary dust particles. Meteoritics Planet. Sci. **31**, 394–402.

This paper describes the first attempt to relate IDPs to D- and P-type asteroids and centaurs by use of spectrophotometry.

Gaffey M. J., Burbine T. H., and Binzel R. P. (1993). Asteroid spectroscopy: progress and perspectives. Meteoritics **28**, 161–187.

A thoughtful review that directly addresses the relationship between asteroid taxonomy and meteorites.

Pieters C. M. and McFadden L. A. (1994). Meteorite and asteroid reflectance spectroscopy: clues to early solar system processes. Ann. Rev. Earth Planet. Sci. **22**, 457–497.

An excellent review paper with especially good sections on recent advances in observational techniques and data interpretation.

Tholen D. J. and Barucci M. A. (1989). Asteroid taxonomy. In *Asteroids II*, University of Arizona, Tucson, AZ, pp. 298–315.

A comprehensive, but rather technical, description of asteroid spectral classification.

THERMAL MODELS, IMPACTS, AND REGOLITHS

Bennett M. E. and McSween H. Y. Jr. (1996). Revised model calculations for the thermal histories of ordinary chondrite parent bodies. Meteoritics Planet. Sci. **31**, 783–792.

The most current thermal models for ordinary chondrite parent asteroids, based on heating by radionuclide decay.

Herbert F. (1989). Primordial electrical induction heating of asteroids. Icarus **78**, 402–410.

An alternative view of asteroid heating, based on electric currents produced by the solar wind.

Keil K., Haack H., and Scott E. R. D. (1995). Catastrophic fragmentation of asteroids: evidence from meteorites. Planet. Space Sci. **42**, 1109–1122.

An excellent summary of the evidence in meteorites for the impact disruption of asteroids and their reassembly into rubble pile structures.

McKay D. S., Swindle T. D., and Greenberg R. (1989). Asteroidal regoliths: what we do not know. In *Asteroids II*, University of Arizona, Tucson, AZ, pp. 617–642.

This chapter provides a thorough description of the lunar regolith and makes comparisons with meteorite regolith breccias.

4 Achondrites

I n early 1980, Mount St. Helens in the state of Washington stirred from a century-long slumber and exploded with a force hundreds of times greater than that of the atomic bombs dropped on Hiroshima and Nagasaki in World War II. Such violent eruptions are among the most awesome spectacles that nature has to offer. Although they conform to the popular conception of volcanic events, these are not the kinds of volcanic eruptions by which most planetary surfaces are fashioned. Less obvious, but volumetrically more important, are the relatively quiet effusions of lavas like those that formed the ocean floors and large portions of some continents on the Earth. These eruptions also have produced massive volcanic shields, such as Mauna Loa in Hawaii, a mountain resting on the seafloor, as tall as Everest but of much greater bulk. The molten materials, called **magmas**, solidify into **basalts**, rocks composed primarily of the minerals pyroxene and plagioclase. Basaltic lava flows likewise have sculpted the face of the Moon and are well represented in rocks returned by *Apollo* astronauts. The occurrence of similar rocks has also been inferred on the surfaces of Mercury, Mars, and Venus. In fact, eruptions of basaltic magmas provide a common thread in the geologic histories of most planetary bodies.

Making liquids from preexisting rocks requires very high temperatures, so magmas commonly form by **partial melting** rather than by complete melting. The rocks remaining after a molten fraction has been extracted are called **residues**. The volcanic edifices constructed from erupted lavas are actually only the surface expressions of vast underground plumbing systems by which magmas rise to the surface. Magmas that congeal at depth produce **plutonic** rocks, whereas those that erupt on the surface form **volcanic** rocks. Like the terrestrial planets, some meteorite parent bodies have experienced igneous activity, and both volcanic and plutonic rocks are represented in meteorite collections. Meteorites that form by crystallization of magma are called **achondrites** (see Figure 4.1), and those that are residues from partial melting are termed **primitive achondrites**.

Figure 4.1: Eucrites and their relatives constitute the most common group of achondrites. This sawed slab of the ALH76005 (Antarctica) eucrite has a brecciated texture produced by impacts into the crust of its parent body. Clasts of igneous rocks, like the one in the middle left section of the slab, indicate that the components of this eucrite breccia originally formed as lava flows. The cube, measuring 1 cm on a side, indicates the scale. Photograph courtesy of NASA.

The igneous origin of achondrites was first recognized by the German petrologist Gustav Rose in 1825. The word *achondrite* means without chondrules, emphasizing its distinction from chondrite, the other major class of stony meteorites. In this chapter we consider a number of different groups of achondrites and primitive achondrites, lumped together into associations that apparently formed on the same parent bodies.

The importance of the study of achondrites (as well as terrestrial igneous rocks) lies in the fact that igneous processes are the primary means by which planetary bodies evolve. Planetary evolution, like its biologic counterpart, generally results in increasing complexity with time. Igneous processes permit planetary **differentiation**, that is, the transformation of an initially homogeneous body into one with a compositionally distinct core, mantle, and crust. Igneous activity also periodically resurfaces planets and transports and releases volatile elements from the interior, sometimes forming an atmosphere or an ocean. Igneous rocks potentially carry records of the compositions of their deep source regions, the identity and the distribution of heat sources, the timing of melting events, the nature of the solidification process, and the internal global characteristics that permitted magma generation and facilitated its movement within the body. These geologic windows are often partly shuttered, so that we can only glimpse these otherwise inaccessible regions, but some achondrites present such interesting views that we may be excused for gawking.

More than You Wanted to Know about Magma

The knowledge of igneous processes gained from several centuries of geologic studies of terrestrial rocks provides such an important foundation for understanding achondrites that it is prudent to review some fundamentals of igneous *petrology* (from the Greek word for rock) before proceeding further. The diversity among terrestrial igneous rocks is greater than among achondrites, so this discussion is restricted to only basaltic rocks and residues that are most relevant to the meteorites.

Basalts are derived by melting of the Earth's mantle, a deep region of **ultramafic** rocks similar in chemical composition to chondrites. Basaltic magmas exist as liquids only at temperatures above 1,000 °C. How are such hot melts generated in the mantle? The heat necessary to melt rocks ultimately derives from the decay of long-lived radioactive isotopes of potassium, uranium, and thorium and from heat generated by crystallization of the Earth's liquid outer core (crystallization releases heat, whereas melting requires an infusion of heat). Temperatures in the planet's interior increase markedly with depth because of the slow rate at which heat escapes to the surface. Rocks are poor conductors of heat, so temperatures in the effectively insulated mantle rise even though heat is generated very slowly. However, increased temperature is not the sole factor that determines whether or not rocks at depth will melt. Pressure also increases with depth, as the mass of overlying material is gravitationally pulled toward the center of the Earth. Increasing the confining pressure generally results in raising the melting point of rocks, thereby counteracting the effect of increasing temperature. It is this pressure increase that keeps the Earth and other planets from having completely molten interiors. It then follows that deeply buried rocks may melt for one of several reasons. First, rocks melt if the temperature is increased to their melting point at the prevailing pressure; this could occur in response to some localized heat source such as a concentration of radioactive isotopes. Alternatively, if the temperature remains fixed, rocks may melt when the pressure on them is lowered. This second case may be a little harder to visualize, but such decompression may occur as plastically deforming but still-solid mantle rock is slowly pushed upward. A third mechanism involves the addition of volatile fluids such as water or carbon dioxide to the rock; an influx of volatiles lowers the rock's melting point so that it spontaneously melts without change in either temperature or pressure. (Most basaltic magmas on the Earth are emplaced at the boundaries between moving tectonic plates, indicating a relationship between tectonics and melting mechanisms. Tectonics does not apply to achondrite parent bodies, so far as we know, but the mechanisms for melting we have just discussed depend on basic physical principles and are likely to be universally applicable.)

Unlike pure substances (such as ice, which melts at $0\,°C$ under 1 atm pressure), rocks consist of mixtures of minerals that melt over a range of temperatures, generally a few hundred degrees. As a rock is gradually heated or the pressure on it is lowered, some combination of minerals will begin to liquify. The first-formed melt will be different in composition from that of the original rock, although the magma composition will steadily approach that of the bulk source rock as melting progresses. However, experience indicates that partial melting produces most magmas, and complete melting is rarely if ever achieved. Despite the fact that magmas form by melting only a modest fraction (commonly less than approximately 25%) of the mantle source materials, it is sometimes possible to reconstruct their mineralogic and chemical compositions. Properties of the source region are imprinted on magmas derived from it, just as the characteristics of certain grapes are passed on to the wines made from them. As in the case of the wine connoisseur, the trick is to develop the skills to recognize these diagnostic properties.

After a certain quantity of partial melt is generated, it may segregate from the solid residue. Because the residue is the unmelted portion, it is better described as a metamorphic rather than an igneous rock, but we will not belabor this technical distinction. Partial-melting residues often *look* like plutonic igneous rocks, even if they are not, and it is sometimes difficult to distinguish the two.

Once free of the enclosing solid residue, a magma begins working its way toward the surface (see Figure 4.2). The driving force behind this upward movement is buoyancy, because magmas are generally less dense than the surrounding mantle rocks. On Earth, the crust is less dense than the mantle and is thought to act as a filter, preventing many denser basaltic magmas from reaching the surface. Only a small fraction of magmas actually erupt; most stall within the crust and solidify as plutons. Successively rising pulses of magma, using the same conduits, form heated pathways that finally traverse the crust and allow volcanic eruptions, in

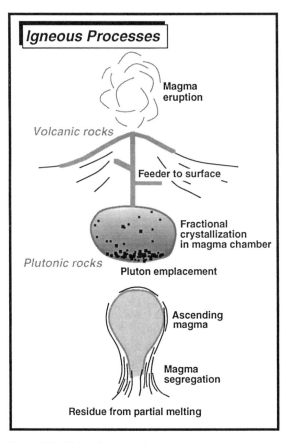

Figure 4.2: This schematic drawing summarizes the important processes that produce igneous rocks. Partial melting of a source region produces magma. Segregation and ascent of magma to higher levels leave behind a solid residue. Some magmas crystallize at depth to form plutons. Most of these undergo some type of fractional crystallization in magma chambers to produce cumulates and fractionated magmas. Such magmas may eventually erupt on the surface as lava flows that solidify rapidly into volcanic rocks.

the same way that chimneys permit smoke to be pulled upward once they are heated. Dissolved gases such as water vapor may be held in solution at depth because of high pressure, but can form bubbles as a magma approaches the surface. This process, analogous to the production of effervescent carbon dioxide as a can of soda pop is opened and depressurized, may accelerate ascent and promote violent eruption.

On cooling, magmas undergo **crystallization** to form igneous rocks. The motions of rapidly vibrating atoms in the melt are damped with lowering temperature until the atoms finally become locked into crystals with orderly structures. Rather than crystallize into a massive crystalline block, a magma will develop numerous crystal nuclei that grow until they impinge on one another. The liquid is eventually transformed into a mass of tightly interlocked grains. The sizes of the crystals are controlled in part by the rate at which the magma cools. Atoms in rapidly chilled magmas quickly lose their motion and combine to form many crystal nuclei. The resulting texture consists of many small grains, sometimes embedded in glass (quenched melt without any crystalline structure), as is typical of volcanic rocks. Deep-seated magmas cool more slowly, permitting atoms to migrate over large distances to the sites of growth and thereby form the larger crystals that are characteristic of plutonic rocks.

Crystallization under plutonic conditions is complicated by the fact that different minerals form sequentially as temperatures decrease during slow cooling. Early formed minerals that are more dense than the surrounding liquid may sink, or they may be carried down by flowing currents, to form concentrations on the floor of the magma chamber. Igneous rocks formed from accumulated crystals are called **cumulates**. The physical segregation of crystals from liquid is called **fractional crystallization**, or fractionation in abbreviated form. (In discussing chondrites, we previously used the term fractionation to allude to the separation of elements with different properties; here, we are separating crystals from liquid.) This process changes the composition of the original magma, by depleting it in the elements that comprise the early crystallizing minerals. Fractional crystallization, a very common process, can happen during transport of magma toward the surface or after its emplacement in a magma chamber. It is important to recognize its effects, because inferences about source regions drawn from highly fractionated magma compositions may be erroneous.

A Geochemistry Lesson

Each of the various igneous processes that lead to the formation of achondrites leaves its own telltale chemical signature on magmas and igneous rocks. Before we can decipher the records of these processes in achondrites, we must digress briefly and discuss some rudiments of geochemistry.

It has already been hinted that some minerals have complex compositions. Olivine crystals, for example, are solid solutions consisting of mixtures of magnesium silicate and iron silicate end members. What has not been mentioned before is that the composition of olivine is dependent on the temperature at which it crystallilzes. Olivine that forms at a high temperature is much more magnesium rich than that which is stable at a low temperature. Pyroxenes behave similarly in terms of their magnesium and iron contents. In plagioclase, the calcium-rich end member is the high-temperature component, and sodium-rich plagioclase forms at lower temperature. A consequence is that accumulations of early formed crystals of olivine and pyroxene from a cooling magma are rich in magnesium, and plagioclase cumulates are rich in calcium. The composition of the complementary magma is, of course, driven to higher iron and sodium contents by fractionation of these high-temperature minerals. This compositional change affords one means of recognizing the products of fractional crystallization. Turning this argument around, we can see that a rock undergoing partial melting will release its lowest melting fraction first. Compared with the solid that is left behind, this melted fraction will be enriched in iron and sodium.

Elements that occur only in trace quantities can be very useful in understanding partial melting and fractional crystallization. For example, the **rare-earth elements** are a group of fifteen elements from lanthanum (La, atomic number 57) to lutetium (Lu, atomic number 71) in the Periodic Table. We focus on these elements as examples of the utility of trace elements, but others may be equally informative. Atoms (actually **ions** – atoms that have gained or lost electrons and thereby acquired an electric charge) of the rare-earth elements are generally trivalent (+3 charge) and are fairly large in size. During crystallization, trace elements must find homes for themselves within ordinary minerals by substituting for other more abundant elements. However, because of their size and charge, almost none of the rare earths fit comfortably into most crystal sites (these and other similarly behaving elements are sometimes called **incompatible elements**, because they are not compatible with common crystal structures). An exception is europium (Eu, atomic number 63), which can form divalent ions (+2 charge) that are approximately the same size and have the same charge as that of calcium ions. Europium is thus tailored to sneak into calcium sites in minerals like calcium-rich plagioclase. This can be seen in Figure 4.3, a plot of the experimentally determined ratios of the concentrations of rare earths in various minerals relative to those of coexisting basaltic magma. The large rare-earth ions are excluded from the small, rigid sites in all these minerals and prefer to remain in the magma (note that all have solid/melt concentration ratios of much less than 1, indicating that they are sequestered into the liquid). However, europium ions are selectively partitioned into plagioclase, creating a spike (the so-called europium anomaly). A few minerals are especially prejudiced against some rare earths while favoring others. The rare-earth elements are plotted from left to right in Figure 4.3 in order of increasing atomic mass. The steeply sloping curve for garnet indicates that this

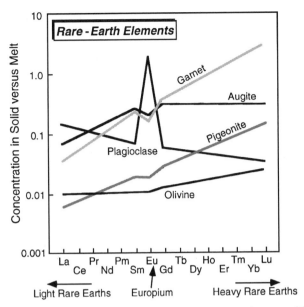

Figure 4.3: Different minerals incorporate trace elements, like the rare earths, into their crystal structures in varying amounts as they crystallize from a magma. The experimentally determined rare-earth-element patterns for five minerals, compared with the abundances of rare earths in the basaltic melt from which they crystallized, are shown in this diagram. The rare earths are plotted from left to right in order of increasing atomic mass, and the vertical concentration scale is logarithmic. Concentration ratios of less than 1.0 indicate higher values in melts than in solids, because all these minerals exclude rare earths from their crystal structures to some degree. Exceptions are plagioclase, which allows europium to replace calcium in its lattice, and garnet, which shows a preference for heavy rare earths over light. Understanding these patterns permits a rock's igneous history to be modeled from its measured trace-element abundances.

mineral prefers the heavy rare earths, because of subtle differences in their sizes.

During partial melting, incompatible elements like the rare earths tend to be flushed from the residue as components of the liquid. Using the information in Figure 4.3 and analyses of rare-earth elements in a basalt, we can model the mineralogy of its source region. During fractional crystallization, the exclusion of rare-earth elements from most minerals causes their concentrations to build up in the remaining magma. As a consequence, cumulates have low abundances and fractionated melts have high abundances of these elements. If plagioclase is one of the minerals that is fractionated, the plagioclase cumulate has extra europium relative to other rare earths (a positive europium anomaly) and the melt has a negative anomaly.

Isotopes also provide information on igneous processes. Radiogenic isotopic chronometers in rocks are reset by melting, so ages derived from them record the times of crystallization for magmas or of partial melting for residues. However, isotopes in slowly cooled igneous rocks can indicate ages younger than these events, because they must cool to their respective blocking temperatures before their clocks are set.

Stable isotopes, like those of oxygen, can be fractionated from each other during melting and crystallization, smearing the compositions of rocks along mass-fractionation lines with slopes of +1/2, as illustrated for terrestrial rocks and several classes of achondrites in Figure 4.4. Because each achondrite class may have its own mass-fractionation line, displaced from the terrestrial line, oxygen isotopes can serve as a means of linking the sometimes diverse igneous rocks formed on a single parent body. However, the fact that rocks from the Earth and the Moon share a common mass-fractionation line indicates that this isotopic signature may not be unique.

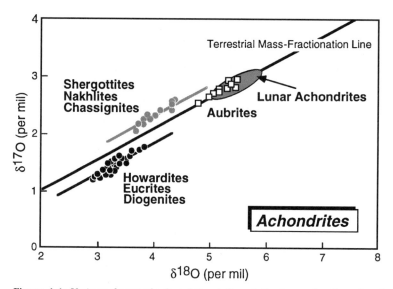

Figure 4.4: Various classes of achondrites define distinct mass-fractionation lines in this diagram of oxygen isotope compositions. The δ notation indicates the ratio of either ^{17}O or ^{18}O to ^{16}O (the most abundant isotope), relative to a standard. Because achondrites cannot move from one fractionation line to another by melting or crystallization, each line in this diagram usually represents a different parent body. However, lunar achondrites and aubrites plot along the terrestrial-fractionation line, indicating that more than one body can have the same oxygen isotopic composition. All analyses were performed by Robert Clayton and Toshika Mayeda (University of Chicago).

Howardite–Eucrite–Diogenite Achondrites

The howardite–eucrite–diogenite (HED) association is the most abundant class of achondrites (approximately 100 meteorites of this group are currently known). It represents a collection of both volcanic and plutonic rocks formed from basaltic magmas. All these diverse meteorites share a common oxygen isotope fractionation line (Figure 4.4) as well as other chemical similarities that mark them as a related suite.

The name **eucrite** comes from the Greek *eukritos*, meaning easily distinguished. The term originally referred to a class of readily recognized volcanic rocks on Earth, but it is no longer used in terrestrial petrology. Eucritic meteorites may be easily distinguished from chondrites, but they look, at least superficially, like terrestrial basalts (see Figure 4.5). Nevertheless, there are some significant mineralogic differences. The plagioclase in eucrites is rich in calcium and contains very little sodium, and the common pyroxene is pigeonite, an iron–magnesium silicate with very little calcium. Terrestrial basalts generally have plagioclase with higher sodium contents and contain augite, a calcium-rich pyroxene, instead of or in addition to pigeonite. Eucrites, like the other members of the HED clan, contain no water, whereas terrestrial basalts commonly contain minor amounts of hydrated minerals or dissolved water in glass. All the iron in eucrites is reduced (either divalent iron in silicates or metallic iron). Terrestrial basalts contain no iron metal and usually have some trivalent iron that forms oxides like magnetite. The implication of these differences is that the source region for eucrite partial melts was distinct in composition and oxidation state from the Earth's mantle, the source region for terrestrial basalts.

The **diogenites** are composed mostly of magnesium-rich, calcium-poor orthopyroxene (similar in composition to the pigeonite in eucrites, but having a different crystal structure caused by slow cooling), with only small amounts of olivine or plagioclase (see Figure 4.5). The high abundance of pyroxenes in diogenites results from fractional crystallization. Diogenitic pyroxenes are more magnesium rich than pyroxenes in eucrites, lending support to the idea that they crystallized from less-fractionated magmas than the eucrites and were concentrated as cumulates.

The conditions under which diogenites and eucrites crystallized can be inferred by a comparison of their textures and compositions. Eucrites have basaltic compositions, indicating that they were once liquids and thus could have erupted onto the surface of their parent body. These are fine-grained rocks, sometimes with small, interlocking crystals that resemble

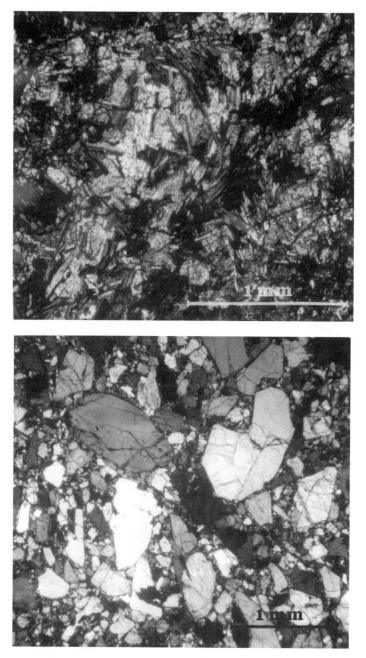

Figure 4.5: In general, the faster a magma cools and crystallizes, the smaller will be the sizes of its crystals. Shown here are microscopic views of thin sections of the Pasamonte (New Mexico) eucrite (top) and the Johnstown (Colorado) diogenite (bottom). Pasamonte contains needles of plagioclase enclosed by pigeonite. The fine-grained texture of the eucrite indicates that it is a volcanic rock. Johnstown consists mostly of orthopyroxene crystals. The diogenite is coarser grained, as appropriate for a plutonic rock, but it has also been heavily brecciated by impact.

those in terrerstrial volcanic flows. Other eucrites consist of equant grains that were apparently recrystallized, that is, thermally metamorphosed. Eucrite layers overlaid and heated by subsequent volcanic flows might experience this kind of annealing, as would basalts buried under blankets of hot ejecta from nearby impact craters. In contrast, diogenites consist of larger, interlocking crystals, as appropriate to plutonic rocks.

There is considerable disagreement about the relationship between eucrites and diogenites. One hypothesis is that diogenites are magnesium-rich pyroxene cumulates of magmas that subsequently fractionated to produce eucrite magmas. A contrasting idea is that eucrites represent partial melts that experienced little or no fractionation. Remelting of the eucrite-depleted residue then produced a second magma that crystallized to form diogenites.

Most eucrites and some diogenites are actually breccias, composed of angular clasts of rock cemented by pulverized mineral grains. Impacts into the parent body that were large enough to excavate plutonic materials were bound to mix eucrite and diogenite fragments. Achondrite regolith breccias containing both eucrite and diogenite clasts are called **howardites**. Plots of the chemical compositions of howardites demonstrate that they lie on simple mixing lines between eucrite and diogenite endmembers.

The crystallization ages of HED meteorites, determined from various radiogenic isotopes, are very old (4.40–4.55 billion years), indicating that the parent body had a relatively short but vigorous igneous history. This suggests an asteroid-sized object, because planets tend to have protracted igneous histories. Thus melting on the HED parent body was an early event in solar system history, roughly contemporary with the thermal metamorphism of chondrite parent bodies. Melting probably utilized the same heat source(s), and this event might plausibly be viewed as metamorphism carried to the extreme.

A connection between HED melting and chondrite metamorphism would be strengthened if it could be shown that the HED parent body was chondritic in composition. Melting of the eucrite source region occurred so early that it seems unlikely that the body had been previously differentiated before this event. Therefore any reconstruction of the composition of the eucrite source region is probably also a fair estimate of the composition of the whole HED parent body. Let us examine how one goes about studying a eucrite to see what minerals comprised its source region.

To illustrate how this might be done, consider a pan of warm water with white crystals, possibly of salt or sugar, on the bottom. If one were given only a sample of water poured from the pan and asked to identify the crystal residue left behind in the pan, how could this be done? One could, of course, taste the water, but, besides being somewhat risky, this might give ambiguous results if the cystals did not impart a distinctive flavor to the water or were a mixture of several substances. Another way would be to cool the water, lowering the solubility of dissolved substances and causing additional white crystals to precipitate. These new crystals might then be identified by other means, such as by their crystal form or chemistry. The crux of this illustration is that the unknown crystals left behind in the pan are the same as those that crystallize from the water on further cooling.

One can study the identity of minerals in the eucrite source region in the laboratory in a somewhat similar manner, by remelting a small meteorite sample in a furnace and directly observing its crystallization during cooling. Liquids that formed by partial melting of a mixture of minerals in a source region should be in **equilibrium** with those same minerals at the rock's melting point. If, at the onset of crystallization, a melt can be shown to form a number of minerals simultaneously, those minerals probably constituted a major part of the source region. Such experiments on eucrites indicate that three minerals – olivine, pyroxene, and plagioclase – all begin to crystallize within a narrow temperature interval of less than $10°$. The implication is that these three minerals were in the residue left behind in the eucrite source region and thus must have been major constituents of the original source rocks. A further important point is that these experiments were carried out at atmospheric pressure. In similar experiments on terrestrial rocks, we can estimate the depth of the source region by adjusting the pressure higher until a number of minerals begin to crystallize simultaneously. The fact that eucrites are in equilibrium with three minerals at low pressure implies that the HED interior was at low confining pressure and thus supports the idea that the parent body was of asteroidal dimensions.

Although experiments suggest that olivine, pyroxene, and plagioclase formed the bulk of the eucrite source region, they do not specify the relative proportions of these minerals. The same liquid could be in equilibrium with a residue mixture containing 99% olivine or only 1%. Luckily, other chemical imprints on magmas can be used to set limits on the proportions of these minerals. The patterns of rare-earth elements in eucrites have been used to estimate that the rock had abundances of olivine, pyroxene, and plagioclase similar to these in chondrites. These calculations suggest that partial melting of a chondritic source rock would yield basaltic magmas similar to the eucrites. Experimental melting of ordinary and carbonaceous chondrites in the laboratory produces similar magmas.

Another group of elements, termed **siderophile** (metal-loving) elements, are depleted in eucrites relative to chondrites. Molybdenum and tungsten are both siderophile and incompatible. If metallic iron is present, they will be dissolved within it, but in the absence of metal, both elements behave like other incompatible elements. During fractional crystallization without metal present, for example, molybdenum and tungsten are partitioned along with lanthanum, a rare-earth element. The observation of constant ratios of these elements in eucrites suggests that metal segregation took place before fractionation of eucritic magmas. Any metal in the original chondritic HED source region may have separated to form a metallic core before eucrite magmas formed and fractionated.

Shergottite–Nakhlite–Chassignite Achondrites

Shergottites take their name from a meteorite that fell in 1865 in Shergotty, located in the Indian state of Bihar. Several distinct kinds of shergottites have been recognized. **Basaltic shergottites** consist primarily of plagioclase and pyroxenes (pigeonite and augite, both showing chemical zoning due to rapid crystallization). The plagioclase is more sodium rich than in HED meteorites, more closely matching that in terrestrial basalts. Shergottites contain abundant iron–titanium oxides (usually including magnetite, indicating that they crystallized under more oxidizing conditions than eucrites). These meteorites are relatively fine grained and apparently formed as volcanic flows or shallow intrusions on their parent body (see Figure 4.6). Their elongated pyroxene crystals commonly have preferred orientations, probably aligned by magma flow. However, the high abundance of the pyroxenes and the presence of magnesium-rich cores within them have been interpreted to indicate that the pyroxene cores are cumulus grains that were added to the basaltic magma before its final solidification. Tiny inclusions of glass, representing magma trapped during crystal growth, occur within the pyroxene cores. Inside these melt inclusions are minute crystals of amphibole, a hydrous mineral whose presence requires that shergottite magmas contain at least a small amount of water. The amphibole is stable at moderately high pressures, so the cores of these pyroxenes may have formed at depth and been carried upward by the ascending magma.

Lherzolitic shergottites are related to the basaltic shergottites, but their coarse grain sizes mark them as plutonic rocks. They are ultramafic cumulates consisting mostly of olivine and orthopyroxene, with minor

Figure 4.6: This image of the QUE94201 (Antarctica) basaltic shergottite illustrates its igneous texture. Many features in meteorites are more readily viewed in backscattered electron images taken with an electron microprobe, and this is one example. The dark grains are maskelynite (a glass formed from plagioclase by shock), the gray grains are pigeonite and augite, and the white grains are oxides and sulfides. Chemical zoning in the pyroxenes is indicated by changes in the gray scale, with darker gray interiors of the crystals being richer in magnesium.

plagioclase and oxides (including chromite, a chromium-rich oxide). The Antarctic meteorite EET79001 (see Fig. 4.7) provides evidence for a clear linkage between the lherzolitic and the basaltic shergottites. This unusual stone contains two basaltic shergottite flows, joined along a planar contact. One of the flows, prosaically called lithology A, is much finer grained than the other (lithology B) and contains xenocrysts of olivine and orthopyroxene. (*Xenocrysts* are foreign crystals, from the Greek word *xeno* for foreign.) These large crystals are disaggregated pieces of a lherzolitic shergottite that was apparently intruded by the basaltic magma and incorporated within it.

All the shergottites have been heavily modified by shock metamorphism. The most noticeable shock effect is that plagioclase has been transformed into maskelynite, a glass produced in the solid state rather than by quenching of a melt. Experiments have produced maskelynite when

Figure 4.7: The EET79001 (Antarctica) meteorite is an unusual shergottite containing samples of two basaltic flows, labeled as lithologies A and B in the sketch. A microscopic thin-section view of the contact between the flows is shown at the bottom of the figure; the location of this section is also shown in the sketch. Lithology A is finer grained than B but contains large xenocrysts (foreign crystals) of olivine and orthopyroxene, apparently disaggregated pieces of lherzolitic shergottite.

Figure 4.8: This microscopic image of the Nakhla (Egypt) meteorite shows large crystals of augite and olivine that accumulated in a magma chamber. The long axis of the photograph is approximately 1 cm.

plagioclase was subjected to shock pressures of at least 30 GPa. Some shergottites also contain veins of shock-melted rock, requiring shock pressures exceeding 45 GPa.

The **nakhlites** are named for Nakhla, a town in Egypt. In 1911 a shower of meteorites rained down on the village, with one stone reportedly striking and killing a dog. These specimens, plus several nakhlites from other localities, are cumulates consisting mostly of calcium-rich pyroxene (augite) and lesser amounts of olivine (see Figure 4.8). A related meteorite is Chassigny (France), the only currently recognized **chassignite**, consisting primarily of olivine, with minor pyroxene and other minerals. Hydrous amphibole, similar to that in basaltic shergottites, has been found in melt inclusions in Chassigny olivine grains. The nakhlites and Chassigny show little evidence for shock metamorphism.

The last member of this unusual group is the Antarctic meteorite ALH-84001, perhaps the most famous meteorite on Earth. Its fame derives from the suggestion that it contains fossilized evidence for early life on Mars (a subject to which we shall return later). ALH84001 is a cumulate rock composed mostly of orthopyroxene. It was initially misclassified as a diogenite, and the mistake was not rectified until a decade after its recovery, when David Mittlefehldt at NASA's Johnson Space Center discovered its true nature while comparing it with members of the diogenite group.

All of these disparate meteorites define a common oxygen isotope fractionation line, as illustrated in Figure 4.4. This line is displaced from both the terrestrial–lunar line and that of HED meteorites. These meteorites also share other geochemical characteristics that indicate they are from the same parent body. Collectively, they are called the SNC meteorites (for the first letters in shergottite–nakhlite–chassignite, and pronounced snik). With the addition of ALH84001, which is sufficiently different from the rest to deserve its own classification, perhaps the acronym should now be revised to SNAC.

The patterns of rare-earth elements in SNC meteorites are compared with those in HED achondrites in Figure 4.9. The ratio of the abundance of each element to that in chondrites is plotted. Chondrites, by definition, have cosmic abundances of rare earths, so this ratio readily shows whether any igneous processes have concentrated or depleted these elements relative to a chondrite precursor. Note that, unlike those of the eucrites, the rare-earth patterns for SNC meteorites are shaped somewhat like a lazy S. Rare-earth patterns in partial melts generally mimic the patterns of their source rocks. If the eucrites formed by partial melting of chondrite, as has

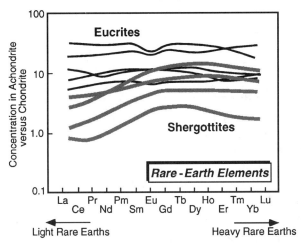

Figure 4.9: Rare-earth elements in rocks and meteorites are normally referenced to the abundances in chondrites so as to highlight fractionations that have occurred during igneous processes. This diagram compares the relatively flat rare-earth patterns in eucrites with the highly fractionated, lazy S-shaped patterns of shergottites. The latter were apparently derived from nonchondritic source regions. The concentration scale is logarithmic.

been argued, we would expect them to have flat rare-earth patterns like those measured. But how do we explain the sinuous shape of the SNC patterns? It is not an easy task to separate rare earths from one another to form such a skewed pattern; clearly the SNC source region was not chondritic in composition. Detailed modeling of this pattern indicates that it could have been produced only by several generations of melting and crystallization, possibly involving the mineral garnet in the source region (recall from Figure 4.3 that garnet discriminates between light and heavy rare earths). Because garnet is a stable mineral only at high pressures, the parent body must have been large. Moreover, the evolution of the SNC parent body must have been more complex than the single-stage melting event envisioned for the HED parent body.

The most unusual features of most SNC meteorites, though, are their young crystallization ages (relative to other meteorites). The magmas that formed the nakhlites and Chassigny crystallized 1.3 billion years ago. That may seem like an ancient age, but it is relatively recent compared with the approximately 4.5 billion-year ages of chondrites and HED achondrites. One basaltic shergottite, QUE94201 (Antarctica), crystallized even more recently at 0.3 billion years ago. The ages of the other shergottites are more difficult to determine, because shock metamorphism has severely disturbed or reset their isotopic clocks. Various radiogenic isotope systems define ages varying between 1.3 billion and 0.2 billion years, and different researchers favor one or the other end of this range. In contrast, the ALH84001 meteorite is very old, apparently having crystallized approximately 4.5 billion years ago. It is clear that igneous activity on the SNC parent body extended over a long period of geologic time, perhaps almost (or even) to the present day.

The view of the SNC parent body that emerges from studies of these meteorites is very different from that of the HED achondrites. It must be a geologically complex body, characterized by multiple periods of igneous activity. Isotopic data suggest that it was differentiated 4.5 billion years ago, and the mantle thus formed had a nonchondritic composition. This source region was remelted approximately 1.3 billion years ago to produce the magmas that formed nakhlites and Chassigny and again more recently to produce shergottite magmas. We cannot reconstruct the exact composition of the source region, because the magmas that produced these meteorites were fractionated during their ascent or emplacement. However, similarities in trace-element abundances and oxidation states, as well as the presence of water, suggest an SNC parent body more closely akin to the Earth than to the HED parent body.

Lunar Meteorites

The scarred face of the Moon has been ravaged repeatedly by massive impacts. Is it possible that lunar rocks could have been ejected by such collisions? Calculations suggest that more than a billion grams of impact ejecta might be lost from the Moon each year and possibly a hundredth of that should be swept up by the Earth. This is only a tiny fraction, less than a percent, of the meteoritic material arriving from space each year, but it is nonetheless significant. If these calculations are correct, lunar rocks should comprise some proportion of our meteorite collections. Such meteorites would be classified as achondrites because of the igneous processes that have dominated the Moon's evolution.

The hypothesis that some meteorites are lunar samples was proposed more than three centuries ago, and it has been revived many times in the ensuing years. The return of lunar samples by *Apollo* astronauts and by unmanned *Luna* landers finally provided a definitive test for this idea. However, from studies of these rocks and soils it quickly became clear that no achondrites had come from the Moon, and in the 1970s this idea once again passed into scientific oblivion.

Then, in 1982, fate intervened. Several American scientists were traversing an ice field near the Allan Hills, Antarctica, on snowmobiles. Their intent was to examine the configuration of the ice sheet; they were not even looking for meteorites. Visibility was poor because of blowing snow, and it was only by the sheerest chance that one of them spotted a small, plum-sized meteorite in his path. The finder, having had considerable experience hunting Antarctic meteorites, recognized that the specimen was different from others he had seen, and he collected it. That evening the weather deteriorated rapidly, making further fieldwork impossible. That chance meteorite was the last specimen collected during the 1981–2 field season in Antarctica.

Some months later a thin section of that meteorite, which had been assigned the name ALH81005, reached the Smithsonian Institution in Washington. The meteorite proved to be an achondrite breccia (see Figure 4.10). In a preliminary description Brian Mason, the Smithsonian's foremost meteoriticist, suggested that "some of the clasts resemble the anorthositic clasts described in lunar rocks." This guarded statement was exciting enough to other meteorite researchers to generate a flurry of sample requests, and by year's end no less than twenty-two research groups were working on this small sample. In early 1983, a special session on ALH81005 was convened to air the results at the annual Lunar and Planetary Science Conference in Houston, Texas. Every participant in the

Figure 4.10: The discovery of lunar achondrites is one of the most exciting revelations of the Antarctic meteorite program. This photograph shows that the ALH81005 meteorite is a breccia containing white anorthosite clasts in a darker matrix. Similar regolith breccias were returned from the Moon by *Apollo* astronauts. The cube measuring 1 cm on a side is for scale. Photograph courtesy of NASA.

session concurred that the meteorite was probably (most said certainly) of lunar origin. Such unanimity of scientific opinion was startling, even allowing for the fact that it was St. Patrick's Day and spirits were high. Apparently at least one meteorite had come from the Moon.

As first noted in the preliminary description, the white clasts in ALH-81005 do resemble some lunar rocks. These are **anorthosites**, cumulate igneous rocks composed almost wholly of plagioclase (in this case, very calcium rich), with only minor olivine or pyroxene. Also present are a few dark clasts of basalt similar to those returned from the Moon. Many of the clasts are recrystallized or even partly melted, and these are mixed with an assortment of broken mineral grains and glass fragments into a brown matrix. The components, plus clumps of glass-bonded aggegates (agglutinates), identify this specimen as a regolith breccia. Similar samples were scattered over the lunar surface at the highlands sites visited by *Apollo* missions.

A more definite indication of lunar origin for this meteorite is provided by its chemistry. For example, the ratios of manganese to iron in ALH81005 resemble those in lunar rocks but are signficantly different from those in terrestrial basalts and in other achondrites, as illustrated in Figure 4.11. The oxygen isotopic composition of this meteorite is also characteristic. ALH81005 plots on the terrestrial mass-fractionation line, as do other lunar samples, but not most other achondrites (Figure 4.4). Finally, this meteorite contains appreciable quantities of the noble gases – helium, neon, argon, krypton, and xenon. These gases were implanted by the solar wind into the matrix of the meteorite while it was in a regolith. In contrast to breccias on the Moon, even the most gas-rich regolith breccias of other achondrite groups have lower concentrations of gases.

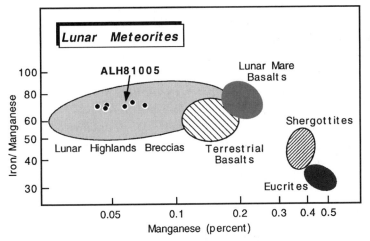

Figure 4.11: Lunar achondrites generally plot in the field of *Apollo* highlands breccias in a diagram of iron/manganese ratio versus manganese abundance. The iron versus manganese abundance is a distinctive property of lunar rocks and serves to distinguish them from terrestrial basalts and other achondrite classes. Lunar achondrite compositions were determined by Herbert Palme and colleagues (Max-Planck-Institut, Mainz, Germany).

Figure 4.12: This microscopic image of a thin section of the Yamato 793169 (Antarctica) achondrite shows a typical lunar basalt, with intergrown pyroxene (gray), plagioclase (white), and iron–titanium oxides and sulfides (black). The image is only 5 mm wide, and the intergrowth texture is typical of volcanic rocks.

Shortly after this revelation that a lunar meteorite had been discovered in Antarctica, Japanese scientists recognized another residing in their Antarctic collection, and more were soon to follow. As of this writing, there are approximately a dozen well-studied lunar meteorites, all but one (an Australian specimen) found in Antarctica. Most, like ALH81005, are breccias composed of materials rich in anorthosite. Four lunar meteorites, however, are basalts or breccias derived from basalts (see Figure 4.12). They consist primarily of pyroxenes (pigeonite and augite) and calcium-rich plagioclase, and are similar in all respects to lunar basalts returned by *Apollo* missions.

Aubrites

In the summer of 1946, a meteorite landed with a splash in a swimming pool at a cattle ranch near Peña Blanca Spring, Texas. Pieces of this stone with a total mass of 70 kg were quickly collected from the bottom of the pool. It would be difficult to envision a more inhospitable landing site for an aubrite. The **aubrites** are achondrites composed primarily of nearly iron-free magnesium pyroxene (enstatite), in contrast to the pyroxene compositions in other igneous meteorites. They also contain small but variable amounts of iron–nickel metal, olivine, troilite, and a variety of exotic minerals, all formed under extremely reducing conditions. These minerals decompose rapidly by reacting with oxygen in water or even air. Most aubrites are marred by brownish spots that result from oxidation of sulfides during their residence on our planet. Although the scientific literature on aubrites contains a number of proposals for their origin by nebular condensation, their textures and other properties clearly indicate an igneous origin.

All the aubrites, except Shallowater (Texas), which is apparently a unique sample, are breccias. Clasts within the breccias are commonly composed of enormous enstatite crystals, some as long as 10 cm and possibly of cumulate origin. The minerals within the clasts show igneous textures, and a variety of rare clasts contain minerals such as plagioclase, augite, and silica, phases not usually ascribed to aubrites. Their rare-earth-element patterns show negative europium anomalies, usually taken as evidence for fractionation of plagioclase from the aubrite magma. The apparent lack of enstatite–plagioclase basalts related to the aubrites has led to a number of suggestions to explain their puzzling absence in meteorite collections. Perhaps the most intriguing is the proposal of Klaus Keil and Lionel Wilson of the University of Hawaii that the basaltic magmas may have erupted explosively as sprays of droplets accelerated by expanding volatiles. If the volatile content were high enough, most of the erupted droplets would have escaped the gravitational hold of the aubrite parent body and have been lost to space.

The oxygen isotopic compositions of aubrites plot along the terrestrial mass-fractionation line (as do the enstatite chondrites), as illustrated in Figure 4.4. The highly reduced nature and mineralogy of aubrites are also very similar to those of enstatite chondrites, suggesting that the aubrite source region was similar to that of these meteorites. The temperatures required for melting enstatite chondrites to form aubrite magma are very high, in the neighborhood of 1,500 °C. Such high temperatures may also be necessary to accomplish the required fractionation of metal from enstatite chondrite starting materials. It is difficult to envision partial melting at such high temperatures, so the aubrite parent body may have melted completely. Consequently, neither the EH nor the EL chondrite group is likely to have been associated with aubrites on the same asteroid.

Some aubrite breccias contain fragments of slowly cooled plutonic rocks as well as melted clasts that formed by impacts and cooled rapidly near the surface. The aubrite parent body may have been collisionally disrupted and reassembled, as proposed for some chondritic asteroids.

Acapulcoites and Lodranites

Acapulcoites take their name from a stone that fell in 1976 on the outskirts of Acapulco (Mexico). **Lodranites** are named for an 1868 fall in Lodran (Pakistan). Although distinct in appearance, the acapulcoites and the lodranites actually form a coherent group with continuously varying characteristics. Together they provide perhaps the best examples of primitive achondrites – meteorites with achondritic textures but with nearly chondritic compositions. There are nearly twenty members of this group, most of which were found in Antarctica.

The acapulcoites and the lodranites share similar mineralogies, both being composed largely of olivine and pyroxene, with minor plagioclase, iron–nickel metal, and troilite. Their silicate mineral compositions contain some oxidized iron and are intermediate between those of E and H chondrites. These primitive achondrites also have similar oxygen isotopic compositions (Figure 4.13), although they do not define a clear mass-fractionation line. This appears to reflect inherent isotopic heterogeneity within the parent body that was not eliminated by igneous processes.

Tim McCoy and his colleagues of the Smithsonian Institution have shown that the acapulcoite–lodranite achondrites represent residues from varying degrees of partial melting of chondrites, ranging from less than 1% to as great as 25%. Acapulcoites show low degrees of melting, evidenced by the presence of veins and small pockets of metal and sulfide. At temperatures of 950 to 1,000 °C, a specific combination (called a **eutectic** mixture) of iron–nickel metal and troilite melted. The silicates in acapulcoites were also recrystallized by metamorphism at these high temperatures, but a few relict chondrules have been observed in several of these meteorites. Lodranites have coarser-grained silicates (see Figure 4.14) and experienced higher temperatures than acapulcoites, generally between 1,050 and 1,200 °C. Eutectic melting of metal and sulfide was nearly complete in the lodranites, and this liquid efficiently drained out. Within this temperature range, silicates also began to melt. Plagioclase in the rock preferentially melted, leaving lodranites depleted in plagioclase relative to the amount in acapulcoites. Although the major element compositions of lodranites are approximately chondritic, their trace-element abundances have been modified noticeably by loss of small amounts of partial melt.

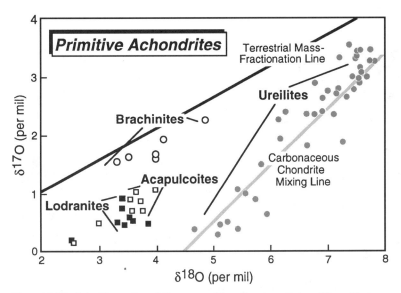

Figure 4.13: Primitive achondrites are residues from partial melting. Their oxygen isotopic compositions preserve the original heterogeneity of their parent bodies, and they do not necessarily define mass-fractionation lines. The acapulcoites and lodranites form a cluster within this diagram, whereas the ureilites scatter along the mixing line defined by refractory inclusions in carbonaceous chondrites. Neither pattern resembles the mass fractionation seen in achondrites that crystallized from basaltic magmas. The brachinite analyses may constitute a mass fractionation line, but scatter in these data makes it poorly defined. The δ notation is as defined in previous oxygen isotope diagrams. All data were determined by Robert Clayton and Toshika Mayeda (University of Chicago).

At least one member of this group, LEW86220, appears to contain excess metal-sulfide and silicate melt. This unique meteorite is an acapulcoite into which metallic and basaltic melts have intruded. Presumably the lodranites were located deeper within the parent body, and rising melts generated from them passed through fractures in the overlying acapulcoites on the way to the surface. Some workers have advocated loss of these melts from the parent body surface as they erupted explosively, in the same manner as that advocated for aubrites.

The timing of melting on the acapulcoite–lodranite parent body has been assessed by a precise lead isotope chronometer. An age of 4.56 billion years has been determined for Acapulco, indicating that temperatures sufficient to cause partial melting must have been achieved shortly after accretion. Ages for argon isotope closure are younger, approximately 4.51 billion years, suggesting that metamorphism continued for tens of millions of years.

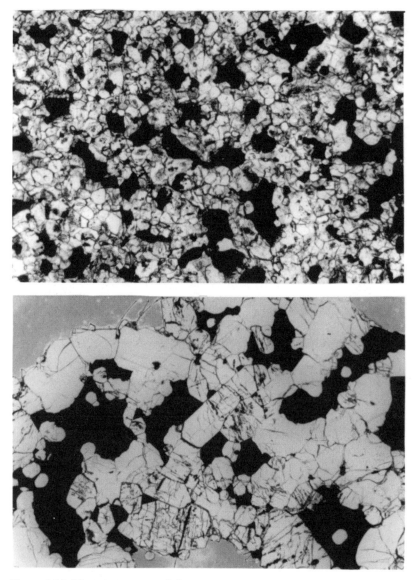

Figure 4.14: These images are of the type specimens of acapulcoite and lodranite meteorites. The photomicrograph (top) shows the fine-grained texture of Acapulco (Mexico), thought to be the residue from which only a tiny amount of melt was extracted. The coarser-grained Lodran (Pakistan) (bottom) has thoroughly recrystallized during extraction of more than 10% melt. Both images, approximately 5 mm across, are courtesy of Timothy McCoy (Smithsonian Institution).

Ureilites

On a September morning in 1886, several meteorites fell near the village of Novo Urei in the Krasnoslobodsk district of Russia. This was a particularly interesting fall for several reasons. One of the stones was soon recovered by local peasants, whereupon it was broken apart and eaten. The motivation for this rather unusual action is not known, but this constituted an impressive feat from a dental perspective, because the meteorite contained numerous small diamonds. The uneaten specimens from this fall proved to be a unique type of achondrite; subsequently recovered meteorites of this class are known as **ureilites**. The forty known ureilites qualify this as the second largest group of achondrites.

The ureilites are arguably the most bizarre and perplexing of all meteorites. They consist principally of the minerals olivine and pyroxene (mostly pigeonite, although augite- and orthopyroxene-bearing ureilites are known). Filling the spaces between the larger silicate grains is a matrix of graphite or diamond (both forms of carbon), iron–nickel metal, troilite, and other minor phases. Reaction with graphite appears to have caused the iron in the rims of olivine and pyroxene crystals to be reduced to metal.

The coarse-grained sizes of ureilites suggest that they formed in the deep interior of their parent body. These crystals typically meet in triple junctions and have curved boundaries. They also show preferred orientations, as illustrated in Figure 4.15. For many years it was disputed whether they formed from accumulations of minerals in a magma chamber or as residues from partial melting. Measurement of the oxygen isotopic compositions of ureilites favors the latter interpretation. These meteorites scatter along the mixing line defined by refractory inclusions in carbonaceous chondrites (Figure 4.13). The oxygen isotope mixing line is considered to be a characteristic of unprocessed nebular materials. In contrast, we have already seen that igneous rocks crystallizing from magmas, such as HEDs or SNCs, plot along mass-fractionation lines. Either the olivine and pyroxene in ureilites never melted, or melting and crystallization occurred on a separate parent body for each ureilite. It seems more probable that the ureilites are unmelted residues left behind after a partial melt was extracted.

This interpretation is not without its problems, however. The rare-earth-element patterns and other chemical characteristics suggest that ureilites are highly fractionated igneous rocks. Perhaps ureilites are residues through which fractionated melts migrated, and some melt was trapped and solidified in the crystalline mesh. The incongruity between

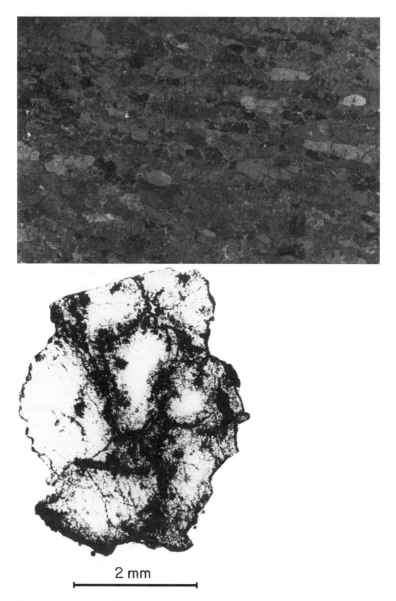

2 mm

Figure 4.15: Two different views of ureilites. The upper photograph shows a polished slab (8 mm long) of the Kenna (New Mexico) ureilite, illustrating a strong preferred orientation of elongated mineral grains. The lower image is a microscopic view of the Haverö (Finland) ureilite. The large white grains are olivine crystals that commonly meet in triple junctions. The dark areas between the grains are composed of finely disseminated metallic iron and intergrowths of graphite and diamond. Photographs courtesy of John Berkley (State University of New York at Fredonia) and Ursula Marvin (Smithsonian Astrophysical Observatory).

igneous and primitive features makes any interpretation of these meteorites problematical.

Another unresolved question is the source of the carbon in ureilites. A few meteorites contain rather large blades of graphite that are sometimes enclosed within the magnesium-rich rims on olivine and pyroxene grains, and in others graphite intrudes the silicates along fractures. Proponents of the cumulate origin favor a purely igneous model in which the carbonaceous matrix represents a crystallized melt trapped between accumulating crystals. Advocates of the residue origin suggest that the carbonaceous matrix may be unrelated to the silicate grains, having been forcefully injected into the rock at some later stage.

The presence of carbon and the oxygen isotope compositions of ureilites suggest a link with carbonaceous chondrites. Moreover, these meteorites contain trapped noble gases in abundances similar to those in carbonaceous chondrites. It seems likely that the ureilite parent body had a carbonaceous chondrite composition. The addition of a small amount of basalt (presumably extracted from these residues) to ureilites can produce a rock with the composition of carbonaceous chondrite.

One of the most interesting characteristics of ureilites (aside from their possible tastiness) is that they have experienced variable but typically intense shock metamorphism. In many specimens, graphite, the original carbon mineral, has been partly transformed by shock into its polymorphs, diamond and londsdalite (polymorphs have the same composition but different crystal structures). Shock has also disturbed the isotopic clocks in ureilites. The olivine and pyroxene assemblage appears to have formed 4.5 billion years ago, but some radiogenic isotopes were redistributed at 4.0 billion years.

Other Achondrites

The list above does not exhaust the recognized classes of achondrites. The **angrites** are a small group of meteorites composed of pyroxene, olivine, and plagioclase. The pyroxene is fassaite, a distinctive composition rich in calcium, aluminum, and titanium. Olivine contains considerable amounts of calcium, and plagioclase is almost pure anorthite (the calcium-rich end member). These mineral compositions reflect the marked enrichment in refractory elements and depletion in volatiles that are characteristic of angrites.

The textures of angrites are variable, but all indicate crystallization from basaltic magmas (see Figure 4.16). They have very ancient ages, approximately 4.56 billion years. The oxygen isotopic compositions of angrites

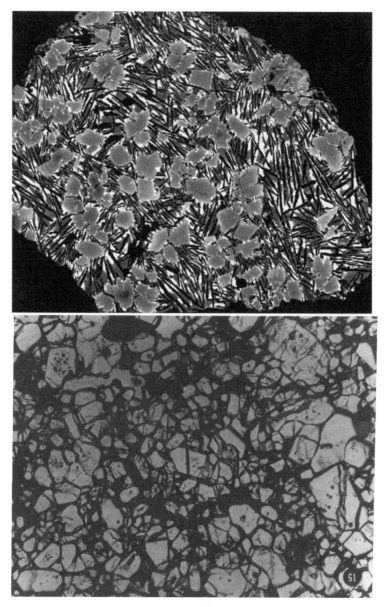

Figure 4.16: The upper photograph shows a backscattered electron image of an angrite, LEW87051 (Antarctica). The large gray crystals are olivine, set in a matrix of plagioclase needles, fassaite pyroxene, and calcium-rich olivine. The width of the image is 5 mm. The lower photograph is a microscopic view of the recrystallized texture of a brachinite from Brachina (Australia). Most of the white grains are olivine, with minor pyroxene. The larger dark interstitial grains are plagioclase. The long axis of this photo is approximately 2.5 mm. Images courtesy of Gordon McKay (NASA Johnson Space Center) and Brian Mason (Smithsonian Institution).

are indistinguishable from those of HED achondrites, but differences in chemistry are interpreted to indicate that these meteorites came from a separate parent body. Experiments suggest that partial melting of chondrites under more oxidizing conditions than those that produced eucrite magmas could have generated angrite melts.

The **brachinites** are a small class of olivine-rich primitive achondrites (see Figure 4.16). Brachina (Australia) resembles Chassigny and was originally classified as a chassignite. However, its ancient 4.5 billion-year age, distinctive trace-element pattern, and oxygen isotopic composition (Figure 4.14) indicate that it and a handful of other subsequently discovered brachinites constitute a distinct group.

Another class of primitive achondrites is the **winonaites**. The oxygen isotopic signatures of the these meteorites distinguish them from brachinites and other achondrite classes. The winonaites appear to be closely related to iron meteorites and are discussed in Chapter 6.

A Personal Touch

In the National Air and Space Museum in Washington, DC, a metal pylon stands just below the Wright brothers' airplane. Embedded in the pylon is a small black rock shaped like an arrowhead. People, young and old, approach it under the eye of a guard. One by one they advance, reach out their hands, touch the rock, and then walk slowly away, smiling or thoughtful. They have just done something that was once impossible. They have touched a piece of the Moon. (From *Getting Our Hands on the Universe* by B. M. French, The Planetary Report, March 1984.)

What is so nicely described here is an intensely human response, understandable even by scientists who handle lunar samples routinely. The touching of something so foreign as a rock from the Moon is an experience not easily forgotten.

Achondrites elict the same kind of wonder (at least in those who know what they are), because they also are the geologic products of other worlds. In this chapter we have discussed many classes of achondrites, the products of melting of preexisting rocks. In addition, we have considered a variety of primitive achondrites, which are the residues left behind when partial melts are extracted from chondrites. Achondrites and primitive achondrites demonstrate that many meteorite parent bodies suffered heating to temperatures greater than those experienced by the parent asteroids of chondritic meteorites.

Despite the fact that most achondrites are very different in composition from terrestrial igneous rocks, they appear to have formed by parallel processes. For this reason, achondrites seem comfortingly familiar to many geologists, at least in comparison with chondrites for which we have no terrestrial analogs. Correctly interpreting the evidence in achondrite associations is a difficult task, plagued by assumptions and fraught with ambiguities. However, the outcome – glimpses into the geologic evolution of other worlds – is certainly worth the effort. In Chapter 5 an attempt is made to define more fully the characteristics of each achondrite parent body and to identify specific bodies where possible.

Suggested Readings

There are very few nontechnical papers on achondrites. Most of the works cited here require some knowledge of the specialized vocabulary of petrology and geochemistry.

GENERAL

Clayton R. N. and Meyeda T. K. (1996). Oxygen isotope studies of achondrites. Geochim. Cosmochim. Acta **60**, 1999–2017.

The figures in this paper are especially useful in defining the oxygen isotope characteristics that are commonly used to distinguish achondrite classes.

Dodd R. T. (1986). *Thunderstones and Shooting Stars: The Meaning of Meteorites*, Harvard U. Press, Cambridge.

This very readable book has an especially good chapter on "When Planets Melt" that describes igneous processes.

McSween H. Y. and Stolper E. M. (1980). Basaltic meteorites. Sci. Am. **242**, 54–63.

A nontechnical paper describing the eucrite and the shergottite meteorites and their possible parent bodies in easily understood terms.

HOWARDITES, EUCRITES, AND DIOGENITES

Hewins R. H. and Newsome H. E. (1988). Igneous activity in the early solar system. In *Meteorites and the Early Solar System*, J. F. Kerridge and M. S. Matthews, eds., University of Arizona, Tucson, AZ, pp. 73–101.

This chapter is difficult reading, but it focuses on igneous processes affecting HED meteorites.

Mittlefehldt D. W. (1994). The genesis of diogenites and HED parent body petrogenesis. Geochim. Cosmochim. Acta **58**, 1537–1552.

A technical paper describing the difficulties in relating diogenites to eucrites.

SHERGOTTITES, NAKHLITES, AND CHASSIGNITES

Floran R. J., Prinz M., Hlava P. F., Keil K., Nehru C., and Hinthorne J. R. (1978). The Chassigny meteorite: a cumulate dunite with hydrous amphibole-bearing melt inclusions. Geochim. Cosmochim. Acta **42**, 1213–1229.

This paper describes Chassigny and reports the discovery of hydrous minerals in it.

Harvey R. P. and McSween H. Y. (1992). Petrogenesis of the nakhlites: Evidence from cumulate mineral zoning. Geochim. Cosmochim. Acta **56**, 1655–1663.

A technical description of the nakhlites, focusing on how their crystallization histories can be used to unravel their origins.

McCoy T. J., Taylor G. J., and Keil K. (1992). Zagami: product of a two-stage magmatic history. Geochim. Cosmochim. Acta **56**, 3571–3582.

An excellent description of the mineralogy, texture, and origin of a basaltic shergottite.

McSween H. Y., Taylor L. A., and Stolper E. M. (1979). Allan Hills 77005: a new meteorite type found in Antarctica. Science **204**, 1201–1203.

This paper provides a description of the first-recognized lherzolitic shergottite.

LUNAR ACHONDRITES

Special issue: The MacAlpine Hills Lunar Meteorite Consortium (1991). Geochim. Cosmochim. Acta **55**, 2999–3180.

A collection of technical papers by many authors, who describe a typical lunar meteorite breccia and its relationship to other lunar rocks and achondrites.

Special issue: Queen Alexandra Range 93069 and other lunar meteorites (1996). Meteoritics Planet. Sci. **31**, 849–924.

An assortment of articles focused on lunar meteorites of basaltic composition.

AUBRITES

Keil K. (1989). Enstatite meteorites and their parent bodies. Meteoritics **24**, 195–208.

This thoughtful review describes the properties of aubrites and presents arguments that they cannot be derived by melting EH or EL chondrites.

UREILITES

Goodrich, C. A. (1992). Ureilites: a critical review. Meteoritics **27**, 327–352.

The incredibly complex and often confusing origin of ureilites is documented in this excellent review article; after reading it, you may not understand ureilites, but you will have been exposed to what is known about them.

ACOPULCOITES, AND LODRANITES

McCoy T. J., Keil K., Clayton R. N., Mayeda T. K., Bogard D. D., Garrison D. H., Huss G. R., Hutcheon I. D., and Wieler R. (1996). A petrologic, chemical and isotopic study of Monument Draw and comparison with other acapulcoites: evidence for formation by incipient partial melting. Geochim. Cosmochim. Acta **60**, 2681–2708.

A thorough description of acapulcoites, documenting their formation by small degrees of partial melting.

McCoy T. J., Keil K., Clayton R. N., Mayeda T. K., Bogard D. D., Garrison D. H., and Wieler R. (1997). A petrologic and isotopic study of lodranites: evidence for early formation as partial melt residues from heterogeneous precursors. Geochim. Cosmochim. Acta **61**, 623–638.

A companion paper to that above, describing the properties and origin of lodranites.

ANGRITES

Mittlefehldt D. W. and Lindstrom M. M. (1990). Geochemistry and genesis of the angrites. Geochim. Cosmochim. Acta **54**, 3209–3218.

A comprehensive study of angrite achondrites.

BRACHINITES

Nehru C. E., Prinz M., Delaney J. S., Dreibus G., Palme E., Spettel B., and Wänke H. (1983). Brachina: a new type of meteorite, not a chassignite. J. Geophys. Res. **88**, B237–B244.

This paper describes the first recognized brachinite.

5 Achondrite Parent Bodies

pollo – the name still evokes a special feeling in anyone old enough to remember sitting riveted to the television to watch this "one giant leap for mankind." The series of missions that culminated in placing men on the Moon ranks as one of the triumphs of human endeavor. The early *Apollo* flights allocated little time for sample collection because of concern about the astronauts' ability to cope with the harsh lunar environment, but on later missions, extended sampling forays, complete with transport vehicles, were carried out with more confidence. A total of 382 kg of lunar rocks and soil were ultimately returned to Earth by the *Apollo* astronauts. Assuming a cost of $24 billion for the total *Apollo* program, this works out to $60,000 per gram of sample. From a scientific perspective, these are bargain-basement prices. Even from a purely economic point of view, the most cynical (but informed) critics must acknowledge that the technological spinoffs have compensated for a large fraction of the cost of this program.

Consider how much more of a bargain are achondrites. Celestial traffic accidents have provided geologically processed samples of other worlds at no cost and with little effort on our part. (Technically, acquiring meteorite samples in Antarctica involves some cost and effort but, relative to spaceflight missions, these samples are cheap and the collection of them is safe.) However, we pay another kind of price in trying to use these meteorites for scientific research: For many achondrites we do not know with certainty from which parent bodies they are derived. Nevertheless, as we see in this chapter, we can make some very informed guesses about the identities of some achondrite parent bodies.

Our Nearest Neighbor

Because the *Apollo* program was mentioned above, let us start with the best-known achondrite parent body, the Moon. The lunar achondrites are clearly exceptions to our generalization about positively identifying achondrite parent bodies, but only because we have *Apollo* samples, as well as those collected on the unmanned *Luna* missions of the former Soviet Union, with which to compare them. What has been learned about the Moon's rocks? The following picture has been pieced together from petrologic, geochemical, and isotopic investigations of lunar samples, as well as from high-resolution photographs and remote-sensing measurements made by orbiting spacecraft.

Even casual inspection of photographs of the near side of the Moon reveals that the lunar landscape can be divided into two kinds of areas. The **highlands** are higher, rougher terrains of lighter color than the **maria** (Latin for "seas"), which are dark, relatively smooth lowlands. The far side of the Moon, which always faces away from the Earth and was never seen until the advent of lunar-orbiting spacecraft, consists almost entirely of highlands terrain. All the *Apollo* and the *Luna* missions landed on the near side, and most sampled the maria, for the obvious reason that smoother surfaces provided safer landing sites. These sampling sites are pointed out in the near-side photograph of Figure 5.1.

The rugged character of the highlands is attributable to ancient meteoroid impacts that formed countless overlapping craters. Some of these craters are huge, spanning hundreds of kilometers from rim to rim. Such intense cratering spread highlands ejecta over much of the lunar surface. Particles of plagioclase-rich rock found in soil collected during the *Apollo 11* mission (a mare landing site and the first sample-return mission) led John Wood of the Smithsonian Astrophysical Observatory and Joseph Smith of the University of Chicago to speculate independently that the highlands consisted of anorthosite. That such rocks are major highlands components was subsequently confirmed by the *Apollo 15* and the *Apollo 16* missions that actually landed in the highlands. Moreover, remote-sensing measurements made by several *Apollo* orbiters (the unlucky astronaut left aloft on each mission actually collected very important data) detected high concentrations of aluminum wherever the spacecraft passed over highlands areas (see Figure 5.2), supporting the idea that they consist of plagioclase-rich (hence aluminum-rich) anorthosites.

At least three compositionally distinctive suites of rocks have been found in the highlands. The **ferroan** (iron-rich) **anorthosites** consist mostly of plagioclase, with only a few percent of pyroxene and olivine. The original plutonic rocks were very coarse grained (see Figure 5.3), although most have been thoroughly granulated and shocked by later impacts. The plagioclase is very calcium rich, and the rocks show no correlation between the composition of plagioclase and the magnesium-to-iron ratios of pyroxene or olivine. The nearly vertical trend of mineral compositions in ferroan anorthosites (see Figure 5.4) was unanticipated,

Figure 5.1: This photograph of the near side of the Moon clearly shows the distinction between the rough, heavily cratered highlands that are light in color and the relatively smooth, dark maria. The arrows mark the landing sites of *Apollo* and *Luna* missions that returned lunar samples to Earth. Photograph courtesy of the Lunar and Planetary Institute.

because it does not follow the normal behavior expected for fractional crystallization. Clasts in lunar highlands meteorite breccias plot, for the most part, within the field of ferroan anorthosites. Radiometric ages for lunar anorthosites are difficult to determine, because of their mineralogical simplicity and resetting of ages by shock, but a few samples are as old as 4.6 billion years. Anorthosites form a crustal layer that is of the order of 50 km thick on the near side and as much as 86 km thick on the far side. This crust formed when the Moon was extensively melted to form a **magma ocean** in its earliest history. The global magma body

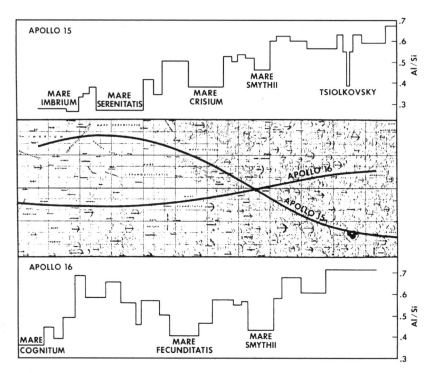

Figure 5.2: The ratios of aluminum to silicon in materials on the lunar surface are shown for two orbiter tracks flown during the *Apollo 15* and the *Apollo 16* missions. The abundances of these elements were determined from x-ray and gamma-ray measurements made on board the orbiting spacecraft and were checked by a comparison of the measurements with analyzed soil samples collected at overflown landing sites. Areas of ancient highlands crust have consistently high aluminum contents. Because plagioclase is rich in aluminum, this experiment supports the idea that the lunar crust is mostly anorthosite. Figure courtesy of NASA.

experienced fractional crystallization as less dense plagioclase crystals floated to the top and solidified to form the feldspar-rich highlands. The difference in thickness of the anorthosite crust on the near and the far sides may be related in some way to the gravitational pull of the Earth acting continuously on one face. The magma ocean stage of lunar history had ended by approximately 4.4 billion years ago.

Soon thereafter, another suite of plutons invaded the anorthosite crust. Samples of these rocks, called the **magnesian suite**, vary in age from 4.4 to 4.2 billion years. They too are cumulate rocks, but they contain, in addition to plagioclase, appreciable quantities of olivine or pyroxene, producing rocks called troctolite and norite, respectively. The magnesian suite exhibits the expected correlation between the calcium content of plagioclase and the magnesium-to- iron ratio of coexisting olivine or pyroxene (Figure 5.4).

Figure 5.3: Much of the lunar highlands consists of coarse-grained anorthositic rocks that formed by floatation of plagioclase crystals in a deep magma ocean. This specimen, composed of white plagioclase grains and dark olivine, formed 4.5 billion years ago and was collected at the *Apollo 16* landing site. It is rare to find such well-preserved pieces of the ancient lunar crust in the rock collections brought back from the Moon, because meteorite bombardment has broken most of them into tiny fragments. The cube is 1 cm on a side. Photograph courtesy of NASA.

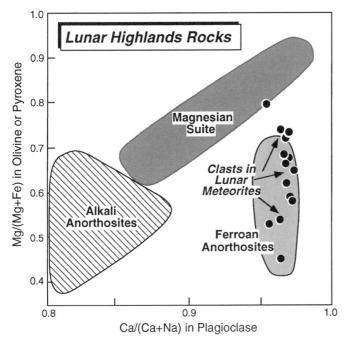

Figure 5.4: Three suites of lunar highlands rocks can be distinguished, based on their mineral compositions. This diagram compares the amounts of magnesium and iron in olivine or pyroxene with the calcium content of coexisting plagioclase. Clasts in meteorites from the lunar highlands plot almost exclusively within the ferroan anorthosite field, suggesting that this suite dominates the highlands, possibly on both sides of the Moon.

Figure 5.5: The lunar maria are vast lakes of crystallized basalt. This sample, collected by *Apollo 15* astronauts, is riddled with holes (vesicles) produced as dissolved gases in the magma formed bubbles during its ascent. The cube is 1 cm on a side. Photograph courtesy of NASA.

A third suite of highlands rocks is the **alkali anorthosites**. To date, these have been found only in the western parts of the Moon's near side. The plagioclase compositions in these rocks are distinctly richer in sodium than those in the ferroan anorthosites (see Figure 5.4), and they clearly formed from different magmas. All the highlands rocks, regardless of the suite to which they belong, are completely free of hydrated minerals and formed under highly reducing conditions.

Shortly following, and perhaps coincident with the formation of the ancient lunar crust, was a stage of cataclysmic bombardment. The outer several kilometers of the Moon were virtually demolished and reduced to a heavily cratered pile of rubble. Battering by large objects stopped rather suddenly at approximately 3.9 billion years ago, but left in their wake enormous circular basins that were the loci for the next (and final) stage of lunar igneous activity.

From approximately 4.0 to 3.2 billion years ago, and perhaps even later, the gigantic basins were filled with vast outpourings of basaltic magma that crystallized to form the maria. These **mare basalts** (see Figure 5.5) consist mostly of pyroxene, plagioclase, and sometimes olivine. Like the highlands rocks, mare basalts contain no water and were formed under highly reducing conditions. Mare basalt magmas are thought to have formed by partial melting of the olivine- and the pyroxene-rich cumulate that settled from the magma ocean as a complement to the anorthosite crust. The evidence for this assertion comes from measured rare-earth-element patterns for these rocks. The highlands rocks have

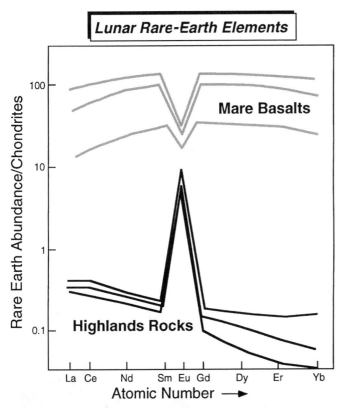

Figure 5.6: Trace elements provide important clues to the origin of lunar rocks. The rare-earth-element patterns for highlands rocks, relative to chondrites, have positive europium (Eu) anomalies (spikes in this diagram), reflecting accumulation of extra plagioclase from the magma ocean. The complementary olivine- and pyroxene-rich cumulate that must lie below the plagioclase-rich crust therefore has a negative europium anomaly. Mare basalts represent partial melts of this underlying cumulate material. As seen in this diagram, they have inherited the negative europium anomaly of their source region.

positive europium anomalies, as illustrated in Figure 5.6, because they consist mostly of cumulate plagioclase that scavanged this element. The plagioclase-depleted cumulate material below must therefore have a negative europium anomaly. Melts derived from this underlying region, the mare basalts, inherited the rare-earth pattern of their parent rocks, as shown in the same figure. Mare basalts vary considerably in their contents of the iron–titanium oxide ilmenite, and titanium contents offer a convenient way to classify these rocks. Experiments indicate that simultaneous crystallization of most minerals occurs at lower pressures for high-titanium basalts than for low-titanium basalts. From such data it has been concluded that low-titanium mare basalts formed by melting at deeper levels than their high-titanium relatives. By approximately 3.2 billion years ago, or slightly more recently, the zone of melting inside the Moon had been drawn down so far and the rigid outer portion of the Moon had become so thick that magmas generated internally could no longer escape to the surface, and volcanism ceased. The absence of mare basalts on the lunar far side probably reflects the greater thickness of anorthosite crust that had to be traversed.

Although fierce bombardment by giant meteoroids ceased approximately 3.9 billion years ago, impacts by smaller objects continued throughout the rest of lunar history at a reduced rate. The effect of this continued battering has been to produce a thick regolith composed of unconsolidated rock dust and compacted breccias like that shown in Figure 5.7. This is the same kind of breccia that comprises most of the known lunar achondrites.

Figure 5.7: Meteoroid impacts have produced a thick regolith on the surface of the Moon. This regolith breccia, returned by *Apollo 17* astronauts, is composed of angular clasts and rock power welded into a coherent rock. This sample is similar in many respects to the breccias formed on asteroid surfaces. Photograph courtesy of NASA.

With so much lunar material collected by the *Apollo* and the *Luna* missions, of what scientific value are a few more small lunar meteorites? The lunar surface area bounded by all the *Apollo* and the *Luna* landing sites is only a small fraction, less than 9%, of the total lunar surface. Odds are high that the lunar achondrites sampled regions outside this limited area, and Paul Warren and Gregory Kallemeyn of the University of California, Los Angeles, have argued that they probably were derived from the far side. Virtually the entire lunar surface has been mapped at eleven wavelengths in the visible and the infrared parts of the spectrum by orbiting spacecraft. A global map of iron content determined from *Clementine* spacecraft data indicates that the highlands sampled by *Apollo* are atypical of the entire Moon's crust. The average composition of the lunar highlands meteorites provides a closer match to this crustal composition than does the average of *Apollo* highlands samples, as shown in Figure 5.8. The analyzed abundances of incompatible trace elements in lunar highlands achondrites, which for the most part cannot be measured from orbit, appear to be consistently lower than those in *Apollo* samples, again suggesting that wider sampling is required for assessing the composition of the lunar crust. In an analogous manner, the *Apollo* mare basalts may not be representative of the maria over the entire globe. The *Apollo*

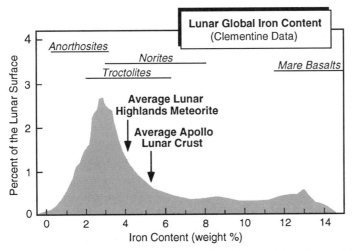

Figure 5.8: The *Clementine* spacecraft used multispectral imaging to determine the global distribution of iron contents of materials on the lunar surface. Shown for comparison are the average compositions of the highlands crust as estimated from *Apollo* rocks and lunar highlands meteorites, as well as bars representing the compositional ranges for various highlands and mare rocks. Lunar meteorites apparently provide a more representative measure of the highlands crust than do *Apollo* samples.

samples are strongly biased toward high-titanium basalts, whereas most orbital measurements and the few known lunar basaltic achondrites are low-titanium varieties.

The last manned lunar mission, *Apollo 17*, returned with its booty of samples in late 1972. That Antarctic snowmobile excursion during which the ALH81005 meteorite was accidentally discovered has been facetiously called the *Apollo 18* mission, because it provided another lunar rock and allowed still more to be recognized. The future recovery of more lunar achondrites, especially in Antarctica, kindles hope that sampling of the Moon will become more complete.

Properties of the Howardite–Eucrite–Diogenite Parent Body

Before we begin our search for the parent body of the HED achondrites, let us briefly review what we have inferred about its structure from the meteorites themselves. Experiments that indicate that eucrite magmas formed at low pressure are consistent with an asteroidal source, because small bodies have low internal pressures. In Chapter 4 we examined evidence that the eucrite source region was chondritic in its chemical and mineralogical composition. Consequently, the bulk of this body (90% or so) must have been composed of olivine and pyroxene, the basic mineral constituents of chondrites. Most of this ultramafic rock was probably highly metamorphosed, and, in places or perhaps throughout, eventually had basaltic melts extracted. Loss of basaltic melt from an ultramafic source rock left a residue consisting mostly of olivine. Eucrite magmas worked their way through this recrystallized interior to the surface. The ascent of many magma pulses was probably halted en route as they were emplaced as plutons that underwent fractional crystallization to form diogenite cumulates. Other eruptions surged onto the surface or were sandwiched between already solidified flows. The gross stratigraphy of the eucrite parent body is thus proposed to be recrystallized ultramafic rock (the residue from partial melting), succeeded outward by diogenite plutons and finally eucrite flows, as illustrated schematically in Figure 5.9. Impacts subsequently blurred this pattern somewhat by mixing surficial eucrite clasts with excavated diogenite to form howardite breccias, at least in the vicinity of large craters. Figure 5.9 also shows a metallic core, which is suggested by the siderophile element abundances in eucrites; however, the existence of a core is unlikely to aid in the search for the HED parent body.

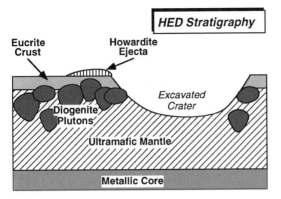

Figure 5.9: The stratigraphy of the outer part of the HED parent body probably looks something like this sketch. This schematic cross section is inferred from the nature of the meteorites themselves. Eucrites are basaltic flows, and diogenites are plutonic rocks that formed at deeper levels. Howardites are mixtures of eucrite and diogenite, excavated in impact events. If the parent body had a chondritic bulk composition, its interior must contain large amounts of recrystallized ultramafic rock (predominantly olivine and pyroxene) that partially melted to produce magmas that formed eucrites and diogenites. The depletion of siderophile elements in eucrites implies formation of a metallic core. The absence of meteorites consisting of recrystallized ultramafic rock suggests that the HED parent body has not suffered catastrophic fragmentation and is still intact.

The dimensions of the interior stratigraphic layers are probably impossible to deduce from meteorite studies, but the thickness of the outside eucrite layer might be estimated from the rate at which the most deeply buried eucrites cooled. The slowest cooling rate naturally corresponds to the most deeply buried rock. Cooling rates have been deduced from features developed within pyroxene crystals. Pyroxenes are complex solid solutions of augite and pigeonite end members at high temperatures, but the homogeneous mixture breaks down as temperature decrease. This unmixing process is called **exsolution**. Like an oil-and-vinegar salad dressing that separates into its two constituents when given sufficient time, the components of pyroxenes unmix under slow cooling. The pyroxenes in eucrites consist mostly of the pigeonite component, so calcium ions diffuse through the solid crystals to certain locations at which augite grows. Augite forms flat plates whose orientation is fixed by the crystal structure of the host pigeonite and whose widths are controlled by the cooling rate. Using the known migration rate of calcium in pyroxene, we can calculate the widths of augite exsolution plates grown at different cooling rates. The slowest cooling rate determined so far is for the Moore County (North Carolina) eucrite. Its cooling rate of 1.6 °C per 10,000 years, representing the most deeply buried of known eucrite samples, corresponds to a depth of 8 km. The eucritic crust must be at least this thick and may be considerably thicker if the body experienced a greater degree of melting.

Howardites contain clasts of both eucrite and howardite, but fragments of the underlying olivine-rich ultramafic zone have only rarely been recognized as components of these regolith breccias. Moreover, olivine-rich achondrites with oxygen isotopic signatures of the HED parent body have not been observed. Impact disruption of the HED parent body would certainly have liberated much more ultramafic rock than eucrite and diogenite. The apparent absence of samples from the deep interior of this body in meteorite collections leads to the conclusion that the HED parent asteroid must still exist.

Looking for a Needle in a Haystack

The mineralogy of eucrites provides a rather distinctive reflectance spectrum with a strong absorption band near 1 μm attributable to pyroxene. As shown in Chapter 4, the spectra of eucrites compare favorably with the spectrum of one particular asteroid, 4 Vesta. The recognition that the spectrum of Vesta was similar to the spectra of eucrites was the first correlation ever made between any asteroid and meteorites by use of this observational technique. Vesta is the third-largest asteroid, with a diameter of approximately 530 km. Its relatively large size probably accounts in some measure for its place of honor as the first asteroid to be linked to a specific kind of meteorite, although its brightness is also due to high reflectivity.

Asteroid 4 Vesta rotates with a period of 5.3 h. Visual images taken by the *Hubble Space Telescope* of Vesta as it rotates (see Figure 5.10) reveal bright areas and dark splotches, similar to those seen on the Moon by the unaided eye. Obviously the surface of this asteroid must be heterogeneous. Spectral reflectance measurements taken during Vesta's rotation have also been used to construct a map of its surface. This is a difficult undertaking, because it is not possible to resolve only a part of the surface. However, the spectrum measured at any one time is integrated over only one hemisphere, the side facing the Earth. When a spectrum is taken at various times during the body's rotation and these spectral snapshots are spliced together, it is possible to make a mosaic map of at least the equatorial region of the asteroid. This amazing map of Vesta is presented in Figure 5.11. Also illustrated is a sketch map of a strip of *Hubble Space Telescope* spectra. Both maps show a predominantly basaltic surface decorated by large regions with different spectral reflectivities. These

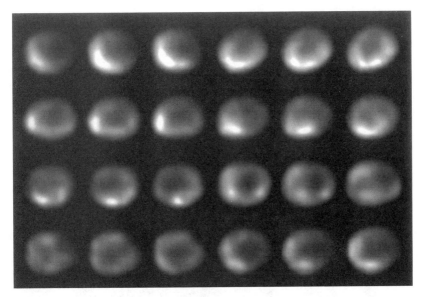

Figure 5.10: The complex surface of asteroid 4 Vesta shows up in these images taken by the *Hubble Space Telescope*. The images, shot over four days in 1994, show one complete rotation of the asteroid (viewed first left to right, then down). The dark splotches may be large craters that excavate mantle materials. Photograph courtesy of NASA.

almost certainly must be impact craters exposing underlying plutonic rocks like diogenite on the crater floors. The olivine spot, which may lie below the diogenite layer, does not correspond to a known meteorite type but is suggestive of the ultramafic mantle material mentioned above. The low-calcium eucrite spot may correspond to a few unusual eucrites, like Bialystok (Poland), which contain pyroxene with very low calcium contents. The observations of Vesta seem to provide elegant confirmation of the parent body stratigraphy inferred from properties of HED achondrites.

In 1997, Peter Thomas of Cornell University and his collaborators published *Hubble Space Telescope* images of Vesta that revealed a huge impact crater, 460 km in diameter and 13 km deep, near the South Pole (see Figure 5.12). This impact event excavated approximately 1% of the asteroid, a volume sufficient to account for a number of small Vesta-like asteroids that orbit nearby. Spectra measured within and around the South Pole crater indicate that rocks on the crater floor are different from those on the surrounding surface and are probably richer in olivine. Thus this impact event probably excavated both the eucrite crust and the underlying diogenite plutons.

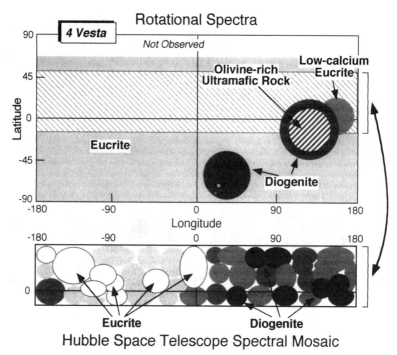

Figure 5.11: The upper map of the equatorial region of the surface of 4 Vesta was made by splicing together individual telescopic spectra taken as the asteroid rotated. Vesta's surface is covered with eucritic basalts. The large circular regions, having spectra similar to those of diogenites or olivine-rich ultramafic rocks, are apparently underlying rocks exposed in the floors of large craters. The strip map at the bottom, whose position is also illustrated in the upper map, is a mosaic constructed from fifty-six *Hubble Space Telescope* images. It also shows that one hemisphere of Vesta is dominated by plutonic rocks similar to diogenites. These maps appear to confirm the stratigraphy of the HED parent body inferred in Figure 5.9. Modified from the published work of Michael Gaffey (Rensselaer Polytechnic Institute) and Richard Binzel (Massachusetts Institute of Technology) and their collaborators.

One rather surprising result of the spectrophotometric survey of the asteroid belt is that no other large V-class asteroids (those with spectra like those of HEDs) other than 4 Vesta have been discovered, along with a handful of small objects in the vicinity of Vesta that appear to be chips dislodged by impacts. It is perhaps misleading to suggest that the search for the HED parent body among the thousands of asteroids in the main belt is like looking for a needle in a haystack. Its distinctive spectrum, high reflectivity, and large size certainly make it stand out from the rest. Despite the fact the 4 Vesta may be unique among surviving asteroids, it remains an amazing achievement that the precise asteroidal parent body for this class of meteorites can be determined with confidence.

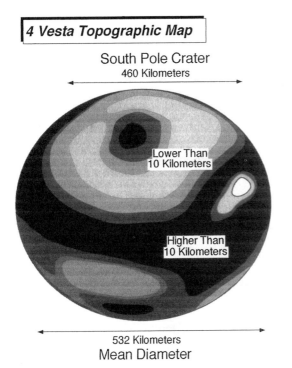

4 Vesta Topographic Map

South Pole Crater
460 Kilometers

Lower Than
10 Kilometers

Higher Than
10 Kilometers

532 Kilometers
Mean Diameter

Figure 5.12: This topographic map of 4 Vesta shows contours of elevation. The bull's-eye structure at the South Pole (shown as the North Pole here because images seen through telescopes are inverted) is a huge impact crater, 460 km wide and 13 km deep. Its central peak is similar to those seen in large craters on other bodies. Excavation of eucrite and diogenite materials from this crater probably produced the HED meteorites. Modified from a projection prepared by Peter Thomas and collaborators (Cornell University).

A Mobile Home for the Aubrites

The near-Earth asteroids are mostly kilometer-sized fragments of main-belt asteroids that have been perturbed into Earth-approaching orbits. Among these nomadic objects is the probable parent body of the aubrites. Asteroid 3103 (it does not yet have a name to accompany its numerical designation) is unique in that it is the only known E-class asteroid within the NEO population. The E-class asteroids have high albedos and relatively featureless spectra (see Figure 5.13), as appropriate for bodies composed of iron-free silicates such as enstatite. Aubrites provide an excellent match for these spectra.

Near-Earth asteroid 3103 is small, approximately 1.5 km in diameter, and elongated in shape. Radar measurements suggest that it has a high

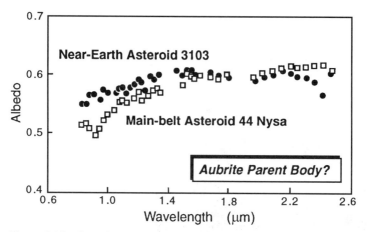

Figure 5.13: The relatively flat spectrum of near-Earth asteroid 3103 is similar to that of 44 Nysa, the largest E-type asteroid, and other E asteroids in the inner main belt. The inferred mineralogies of these bodies are similar to those of aubrites. The spectra were obtained by Michael Gaffey and his colleagues (Rensselaer Polytechnic Institute).

degree of surface roughness, possibly consistent with a brecciated rubble-pile structure. Obviously this asteroid is far too small to represent more than a fragment of the original aubrite parent body. Asteroid 3103 is merely a mobile home of sorts, a transient body from which smaller meteorite chunks were spalled off by later impacts. Fragments knocked off an Earth-crossing asteroid are much more likely to encounter the Earth before they are otherwise lost than are pieces liberated from a typical main-belt asteroid.

The elliptical orbit of asteroid 3103 extends from inside the orbit of the Earth all the way to the asteroid belt, with an aphelion (1.9 AU) in the innermost part of the main belt. This locale, called the Hungarias, is the home of nearly all the known E-type asteroids in the solar system. To make the connection even stronger, the orbits of Hungaria asteroids are tilted approximately 28° from the ecliptic plane, as is the orbit of asteroid 3103. Thus it seems probable that the original trailer park for near-Earth asteroid 3103 was the Hungaria region. The asteroids that populate this region are small, with an average diameter of only approximately 6 km, so they too probably are collisional fragments of some larger body.

Whence Primitive Achondrites?

Specific parent bodies for the various classes of primitive achondrites have not been identified, but we can make some inferences about their properties. Not surprisingly, the spectra of acapulcoites are similar to those of ordinary chondrites and thus to those of S(IV) asteroids. However, extraction of melts to produce lodranite residues in the body's interior would result in rocks depleted in plagioclase, metal, and sulfide, and containing silicates with higher iron contents. Impacts into this object might expose lodranite by stripping off the basalt or acapulcoite cover. Lodranites appear to be spectrally similar to a variety of S-asteroid subtypes [suggestions include S(III), S(IV), and S(V), depending on the amount of melt extracted]. The spectrum of a rubble-pile asteroid composed of acapulcoite and lodranite fragments, as well as remnants of the basaltic crust, would be almost undecipherable.

Ureilites were derived from a body with a particularly complex history. The interior was partially melted and basaltic magma was extracted. The ureilite residue was then held at temperatures of at least 1,250 °C for some time, causing it to recrystallize. The reduction of iron seen in the rims of silicate grains suggests reaction with graphite, a reaction that is extremely sensitive to pressure. Estimates of the pressure necessary to account for the compositions of reduced silicates correspond to depths of approximately 100 km, possibly a minimum size for the ureilite parent asteroid. Various workers have called on explosive volcanism, shock heating by impact, or impact of a carbonaceous asteroid into an already molten body to explain the curious characteristics of these meteorites. Because of their high carbon contents, ureilites are very dark, with albedos of only approximately 7% . The spectra of low-albedo members of the S(I), S(II), and S(III) subtypes might conceivably correspond to the mineralogy of ureilites, but no convincing linkages have yet been documented.

Judging from their spectra, we can see that asteroids of the A type and the S(I) subtype are dominated by olivine and S(II) asteroids additionally contain a small amount of calcium-rich pyroxene. Either of these subclasses might correspond to the mineralogy of brachinites.

The S-type asteroids are thus interpreted to represent potentially a range of meteorite types, from ordinary chondrites to primitive achondrites derived from them by small to modest degrees of melting. The diversity within the S-asteroid class is probably the direct result of the igneous processes that many of these asteroids experienced.

Melted Asteroids

Chondrites represent primitive solar system materials that have largely been spared the effects of geologic processing, whereas achondrites have not. Earlier we learned that asteroids with the spectral properties of various

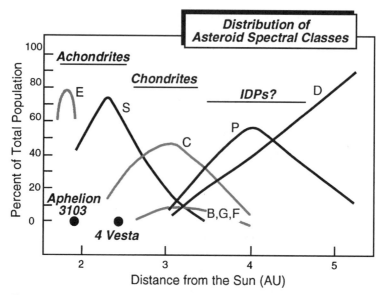

Figure 5.14: Each asteroid spectral class has a preferred location within the main belt. Aubrite parent bodies are probably represented by E-type asteroids, whose location corresponds to the aphelion of near-Earth asteroid 3103. The location of 4 Vesta, the HED parent body, is also illustrated. Other achondrite parent bodies are probably represented by some subtypes of S asteroids.

chondrite groups and possibly interplanetary dust particles (IDPs) occupy preferred regions within the main belt. Let us now see how they compare with the locations of achondrite parent bodies (see Figure 5.14). 4 Vesta, the only large V-type asteroid, is located at a mean heliocentric distance of 2.4 AU. Its location is similar to that of the S asteroids, many of which may be parent bodies for some types of achondrites or primitive achondrites. Near-Earth asteroid 3103, inferred to be the parent body of the aubrites, was apparently derived from the inner asteroid belt. It is spectrally similar to the E-type asteroids, which occupy a region inside the S asteroid swarm. The aphelion of asteroid 3103's orbit is located within the E-asteroid region. Thus, as far as we can tell, all achondrite parent asteroids lie inside a ring located approximately 2.7 AU from the Sun. Not all bodies inside this distance were necessarily melted (the highly metamorphosed ordinary chondrite parent bodies may reside near the outside edge) but all were heated to high temperatures. Asteroids between 2.7 and 3.4 AU may be dominated by carbonaceous chondrite parent bodies that have suffered aqueous alteration. Beyond 3.4 AU, ices appear to have been preserved, and these bodies may be sampled only as IDPs.

Let us make the assumption that the heat source for either partial melting on achondrite parent bodies or metamorphism or aqueous alteration on chondrite parent bodies was the same, which is tantamount to saying that igneous activity is simply an extension of metamorphic processes. Several heat sources have previously been suggested. Electromagnetic induction heating bears an obvious relationship to heliocentric distance, because the intensity of the solar wind that would cause it probably decreased with distance from the Sun. Highly simplified models of induction heating have a number of unspecified parameters – the duration and the intensity of the solar wind and asteroid electric conductivity – allowing almost any result one chooses. Moreover, newer models of T Tauri stars suggest that much of their mass is ejected from their polar regions rather than out through the disk-shaped nebula.

Radioactive decay of ^{26}Al has also been invoked to explain the observed pattern of asteroid thermal histories. During the period of asteroid accretion, ^{26}Al was rapidly undergoing decay. Because the rate of a body's accretion depends on the density of nebular matter available in its vicinity, objects in the denser inner portion of the asteroid belt would presumably have accreted earlier. Therefore bodies in the inner belt would have incorporated greater amounts of live ^{26}Al and thus reached higher temperatures. This is illustrated quantitatively in Figure 5.15, a plot of the diameters of asteroids versus their locations within the main belt. The curves in this figure are contours of the maximum temperatures achieved because of the decay of ^{26}Al, with accretion times (in millions of years after the formation of refractory inclusions, the oldest known nebular objects) shown above the box. The vertical shaded bars superimposed on the diagram separate regions where silicates would begin to melt (requiring temperatures above approximately 1,100 °C), ice would melt (above 0 °C), and ice would be preserved (below 0 °C). These regions correspond to the portions of the asteroid belt inferred to contain the parent bodies of achondrites, altered carbonaceous chondrites, and IDPs, respectively. The diagram shows that 100-km-diameter asteroids would experience the appropriate thermal histories if they accreted within a few million years after formation of the nebula. This calculation demonstrates that ^{26}Al is a plausible heat source, provided that it was widely and uniformly distributed in the nebula.

The HED parent body presents a special challenge to the idea that asteroids were heated by ^{26}Al decay, because the ages of diogenites and some eucrites suggest that the body was hot for perhaps 100 million years, much longer than the effective lifetime of ^{26}Al. By using constraints provided by HED meteorites and 4 Vesta, we can calculate the thermal history of this unusual asteroid. This is a much more complex task than determining the thermal histories of chondrite parent bodies, because the production and the movement of silicate magmas and the separation of a

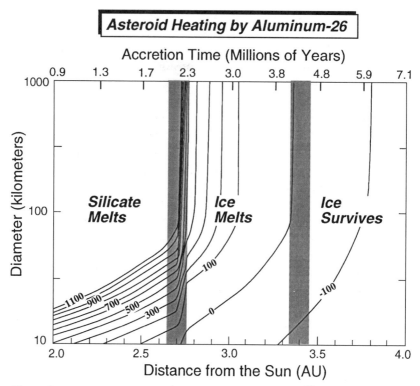

Figure 5.15: Asteroid heating by rapid decay of radioactive [26]Al can quantitatively explain the thermal histories of bodies in the main belt. The maximum temperatures (in degrees centigrade) achieved in asteroids of various sizes are shown by numbered contours. Vertical shaded bars divide the main belt into regions in which silicates melted (requiring temperatures of at least 1,100 °C), ice melted (requiring temperatures above 0 °C), and ice was preserved. If the rate of accretion was proportional to the density of material in the nebula, bodies closer to the Sun would have accreted earlier (and thus contained a higher proportion of live [26]Al) than those at greater distances. The thermal contours correspond to bodies that accreted at times shown across the top of the diagram, in millions of years after the formation of refractory inclusions (the oldest solar system objects). Planetestimals with diameters of 100 km or greater experienced the required temperatures at the appropriate heliocentric distances. Calculations were performed by Robert Grimm (Arizona State University) and the author.

core are complicated events that alter how heat is generated and lost. In one model, core formation began approximately 2 million years after accretion, followed by melting of silicates to produce eucrites. Continued melting of this source region then produced magmas that underwent fractional crystallization to produce diogenites. It is also possible that Vesta experienced such high temperatures that it formed a magma ocean, like that on the Moon. The partitioning of [26]Al into eucrite magmas during melting caused the crust to become very hot, effectively insulating the mantle and preventing heat from escaping from the deep interior. Vesta lost heat so slowly that parts of its mantle remained molten long after the short-lived ^{26}Al was exhausted, accounting for the general absence in eucrites of detectable ^{26}Mg, the decay product of ^{26}Al.

On a Grander Scale

The success of spectrophotometry in identifying the HED parent body leads to the expectation that finding the SNC parent body should be an equally easy task. The reflectance spectra for most asteroids as bright as Vesta down to a few kilometers in diameter have now been measured, but no respectable match has been found for the shergottites and their relatives. Could it be that the SNC meteorites were not derived from an asteroid?

One of the loosely coded but obvious patterns of biology is that larger members of a species (say, elephants and redwood trees) have lived longer than smaller members. If a solar system body can be construed as having a lifetime, it would be based on the length of time the body can sustain internal heat, reflected in the duration of igneous activity. It is interesting that, as in the biological analogy, these geologic lifetimes also show a rough correlation with body sizes, as illustrated in Figure 5.16. Igneous processes on asteroid 4 Vesta ceased approximately 4.40 billion years ago, halting soon after they had begun. Lunar volcanism persisted until approximately 3.2 billion years ago, and volcanic plains on Mercury may be as young as 2.5 billion years. Martian igneous activity is thought to have lasted at least until 1.1 billion years ago and probably to more recent times. The Earth and possibly Venus are still volcanically active. The one known exception to this relationship between planetary size and duration of volcanism is Jupiter's satellite Io, a currently volcanically active body only slightly larger than the Earth's Moon. The reason for this discrepancy is discussed below.

It may seem unclear how the ages of planetary volcanic features, from which we have no samples to date radiometrically, have been determined. We measure these by counting the number of craters per unit area in photographs of uneroded volcanic surfaces; the older a surface is, the more impacts it will have accrued. If the cratering rate is known or can be estimated, an absolute age can be determined from the crater density. The production rate of craters has been calibrated for the Moon, because we have radiometrically dated samples from the same volcanic terrains for which crater densities have been counted. The meteoroid flux apparently

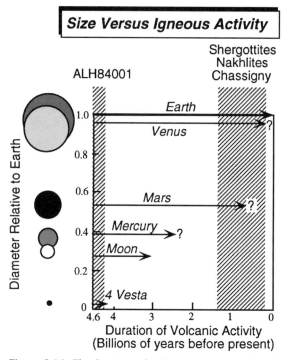

Figure 5.16: The duration of volcanic activity is related to planet size. In this figure, planetary diameters are shown as fractions of the Earth's diameter. The length of time from the beginning of the solar system until volcanism ceased is illustrated by arrows. Question marks for most planets reflect the uncertainty in ages of volcanic plains determined from measured crater densities. The HED parent body, 4 Vesta, supported igneous activity for only a short time after its formation. Larger bodies sustained igneous activity for longer periods, because they retained heat generated by radioactive decay of long-lived radionuclides. The crosshatched areas represent the crystallization of ages of various SNC meteorites. The long duration of igneous activity on the SNC parent body implies that it is a large planet.

was far from being linear with time, because of the heavy bombardment experienced before approximately 3.9 billion years ago. Therefore the ages of surfaces must be estimated from graphs that show how crater production rates have changed with time. The cratering rates for other planets are not known precisely, but can be estimated from probabilities of collisions with other bodies versus the Moon.

The cause of the relationship between planetary size and duration of volcanism is easy enough to fathom. Volcanism is driven by internal heat, generated by radioactive decay of both short- and long-lived radionuclides and possibly by core crystallization. Rocks act as thermal insulation that holds in the heat, and the thicker the rocky blanket, the longer heat can be retained internally. Small bodies may experience heating for a short time because of ^{26}Al decay, which generates heat faster than it can be conducted away, but they cannot retain the more gradually produced heat from fission of the long-lived radionuclides of uranium, thorium, and potassium. Larger planets, being better insulated, can utilize these lumbering decay schemes to fuel their internal heat engines for billions of years. The explanation of the current volcanic activity on the small Jovian moon Io is that its heat is not internally derived from radioactive decay, but rather is due to a peculiar and continuing gravitational flexing by massive Jupiter.

The reason for this discussion of geologic lifetimes is what it portends for the SNC meteorites. Remember that these meteorites have crystallization ages that span the period from 4.5 billion to 0.3 billion years ago and probably even younger. Comparison of these ages with the duration of igneous activity for various solar system bodies, as shown in Figure 5.16, suggests that the SNC meteorites must be derived from a large planet. However, the problems associated with ejecting rocks from the surface of a planetary body are formidable.

One possible way out of this dilemma is to ascribe the origin of SNC magmas to impact melting on an asteroidal body. A small amount of melting of target rocks is expected during large impact events. However, this is probably not a satisfactory solution. The absence of unmelted rock clasts and the presence of cumulate textures in these meteorites make them very different from impact melt rocks typically found in terrestrial and lunar craters. Differences in the initial strontium isotopic compositions of individual meteorites are also inconsistent with this hypothesis, because impact melt sheets appear to be fairly well homogenized.

The evolution of the SNC parent body inferred from study of these meteorites is also consistent with a planetary source rather than with an asteroidal source. The isotopic and trace-element data for SNC achondrites suggest that a complex sequence of igneous events was necessary to produce these rocks. Planetary differentiation occurred approximately 4.5 billion years ago, at which time a core, a nonchondritic mantle, and a crust formed. ALH84001 is a sample of this ancient crust. Partial melting of the mantle source region followed at approximately 1.3 billion years ago, and fractionation of this magma produced the nakhlites and Chassigny. Later remelting of the mantle produced the shergottite magmas. What emerges is a picture of a geologically active world, compositionally similar in many respects to the Earth and capable of multiple episodes of igneous activity that spanned most of geologic time. Such complex behavior is very different from that thought to be possible on asteroids.

The separation of light and heavy rare-earth elements in SNC meteorites is so extreme as to require several igneous events for accomplishing it. This lends credence to the complex evolution already inferred. Very few minerals are capable of discriminating between light and heavy rare earths; most minerals just reject them all. One notable exception is garnet, which has a strong preference for heavy rare earths and is intolerant of the lighter ones. For example, the concentration of ytterbium, one of the heaviest rare earths, in garnet is typically at least 100 times that of lanthanum, the lightest rare earth, relative to their chondritic abundances. One explanation proffered for the highly skewed rare-earth patterns in SNC meteorites is that garnet was a mineral constituent of their source region. The rare earths would not have been distributed equally among the various mantle minerals in this case, and partial melting would have released different amounts of light and heavy elements. Garnet, an important mineral in terrestrial mantle rocks, is a dense phase that forms in ultramafic rocks only at high pressures. The internal pressures in small bodies are so low that garnet is not stable even at the centers of the largest asteroids.

The Red Planet

There is considerable evidence suggesting that the SNC parent body is a planet. The difficulty in removing rock samples from a planetary gravity field demands that we give preference to the smallest suitable planet we can find. The young crystallization ages of most of these achondrites appear to limit our consideration to Mars. However, the proposal that SNC meteorites could be Martian rocks initially met with general disbelief, if not downright hostility. This understandable response stemmed largely from the fact that no lunar achondrites were then known. The logic went something like this: If meteorites have not been ejected from the surface of the Moon, which is smaller and has a lesser gravitational field than Mars, how could they escape from a larger body? The discovery in Antarctica of lunar achondrites soon squelched this particular argument, although finding a mechanism to launch rocks from Mars remained a contentious subject.

Mariner and *Viking* photographs of the Martian surface in the 1970s disclosed the existence of an ancient, heavily cratered southern hemisphere and a northern hemisphere dominated by younger volcanic plains and gigantic volcanoes (see Figure 5.17). Some of the volcanoes appear to be quite young and might be suitable sources for shergottite and nakhlite magmas, but such young terrains occupy only a small part of the Martian

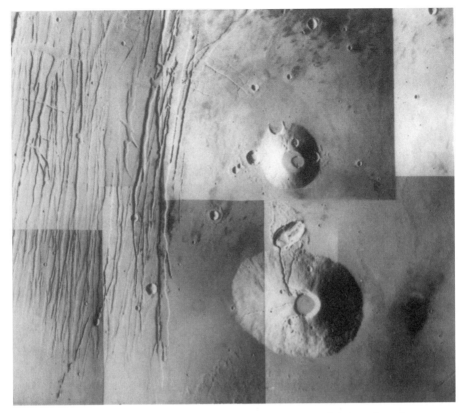

Figure 5.17: The volcanoes in this *Viking* photograph are in the Tharsis region, the most prominent center of recent igneous activity on Mars. Uranium Tholus (top) is 60 km in diameter, and Ceranius Tholus (bottom) is 120 km across. The large oblong crater just north of Ceranium Tholus has been suggested as a possible launching site for some SNC meteorites. The fractures at the left are part of a set of stress features that were produced by bulging in this region, possibly because of ascending magmas. Photograph courtesy of NASA.

surface. It is surprising that we have so many very young SNC meteorites. Similarly, the existence of a 4.5 billion-year-old meteorite among this collection is unexpected. Most of the Martian highlands are thought to have formed at the end of the heavy bombardment that affected the Moon, implying that they are perhaps 4.0 billion years old. The only ancient SNC meteorite, ALH84001, is significantly older than this.

When the two *Viking* spacecraft landed on Mars, they scooped up small shovelfuls of soil and performed quantitative chemical analyses. Similarly, the *Mars Pathfinder* rover analyzed soils at its landing site. These samples provide an intriguing comparison with shergottites. At all three landing sites, the regolith was found to contain high quantities of chlorine and sulfur. This enrichment in volatile elements has generally been interpreted to mean that the uppermost soil is not simply powdered igneous rock, but has had chlorine and sulfur added, either from volcanic exhalations or by interaction with groundwater. However, the average chemical composition of the soil, when extra chlorine and sulfur are removed, is similar to the analyzed composition of basaltic shergottites, but not to other known kinds of achondrites.

When viewed through a telescope, Mars is seen to have bright red regions, thought to be coated with dust, and dark gray regions, probably combinations of soil and bedrock. Reflectance spectra of the dark regions show two prominent absorption bands that indicate the presence of pyroxenes with compositions like those in basaltic shergottites. This observation reinforces the conclusion that at least portions of the Martian surface may be covered by volcanic plains of shergottite. Spectra corresponding to plutonic rocks like nakhlites and Chassigny have not been reported, but higher-resolution imaging from Mars orbiters may eventually reveal the locations of such rocks on the surface. The compositions of rocks at the *Mars Pathfinder* landing site are distinct from those of SNC meteorites, but some geochemical peculiarities of these meteorites are also seen in these rocks (see Figure 5.18).

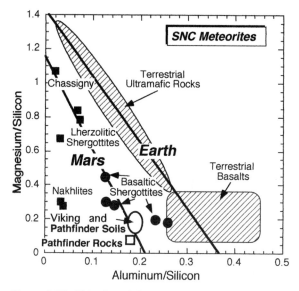

Figure 5.18: This plot of element ratios illustrates that the peculiar compositions of SNC meteorites are also seen in rocks and soils analyzed by the *Viking* landers and the *Mars Pathfinder* rover. Meteorite analyses by Heinrich Wänke and Gerlind Dreibus (Max-Planck-Institut, Mainz, Germany).

If SNC meteorites were blasted by impacts off the Martian surface, it seems likely that they should have been shocked in the process. Indeed, the shergottites are among the most severely shocked meteorites, but the nakhlites and Chassigny show only minor shock effects. Several shergottites even contain tiny pockets and veins of impact-melted glass, the ultimate response of rocks to shock. These little glassy beads have provided the most definitive link with the red planet. This fascinating discovery unfolded almost by accident. In 1982 Donald Bogard and his colleagues at NASA's Johnson Space Center analyzed gas released by heating separated glass fragments in an Antarctic shergottite, EET79001. One of the isotopes of gaseous argon, ^{40}Ar, is the decay product of radioactive ^{40}K and thus forms the basis of a radiometric dating technique. From this experiment, these scientists expected to determine the timing of the shock event during which the melt glasses formed, but the age they obtained (in excess of 6 billion years!) was absurd. They correctly reasoned that this glass must contain some extra ^{40}Ar, unrelated to the decay of ^{40}K in the sample. This gaseous isotope was trapped at the time of shock melting, presumably

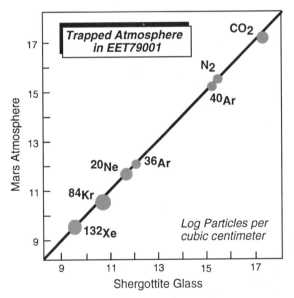

Figure 5.19: Trapped gases in the EET79001 Antarctic shergottite link the SNC meteorites to Mars. Pockets and veins of shock melt in this meteorite formed by impact, which also implanted atmospheric gases in the liquid before it solidified as glass. The abundances of carbon dioxide, nitrogen, and various nonradiogenic isotopes of gaseous argon, neon, krypton, and xenon in the glass are identical to those measured in the Martian atmosphere by *Viking* spacecraft. These measurements were summarized by Robert Pepin (University of Minnesota).

from an atmosphere on the SNC parent body. The measured abundances and the isotopic compositions of argon, krypton, xenon, and nitrogen, as well as the amount of carbon dioxide in the trapped-gas component, were subsequently determined. What makes this interesting is that the trapped gas in the meteorite closely matches the composition of the Martian atmosphere measured by Viking landers, as shown in Figure 5.19. The Martian atmospheric composition is unique, as far as we know, and its incorporation into this meteorite is thought to serve as a diagnostic fingerprint for its parent body. Similar trapped gases have now been analyzed in several other SNC meteorites.

Figure 5.20: This view of the Martian surface shows vesicular volcanic rocks littering the ground. Could these be shergottites? The metallic object is the Viking lander's soil collector arm. Photograph courtesy of NASA.

The recognition that we have samples from Mars (see Figure 5.20) has led to a greatly improved understanding of the geology of that planet. The chemical compositions of the Martian mantle and core have been estimated from SNC meteorites, and, with that information, it is possible to constrain the mineralogy of the interior. Radiogenic isotopes in these meteorites indicate that planetary differentiation occurred approximately 4.5 billion years ago. The SNCs are very slightly magnetized, suggesting the possible presence of a weak magnetic field in the past. The amount of water in Martian basaltic magmas has also been estimated based on that required for stabilizing amphibole. The importance of this estimate lies in the fact that water is delivered to the surface by volcanism. The amount of water outgassed from the Martian interior was apparently rather modest, compared with that of the Earth, and its hydrogen isotopic composition indicates recycling of water through the atmosphere and underground reservoirs. The existence of small amounts of weathering products in SNC meteorites, mostly clays, carbonates, and salts, provides information on soil-formation processes. And, most intriguing of all, is the suggestion that at least one of these meteorites may contain evidence of early Martian life (a subject to be addressed in a later chapter).

Melted Clues

A medical practitioner must use all the indirect information that can be gleaned from symptoms and relatively harmless tests to dioagnose a patient's illness. There may be more straightforward ways to find out what is wrong, but invasive tests and direct observation might endanger the patient's life. Likewise, for students of the Earth's interior, no indirect information can be wasted, because the region under scrutiny is not directly accessible. Almost any kind of natural or manmade vibration must be analyzed for its trajectory and travel time to yield data on interior composition and heterogeneity. Natural force fields must be measured and modeled. Occasional nodules of mantle and deep crustal rocks accidentally transported to the surface by ascending lavas must be chemically and petrologically dissected. And finally, the information from magmas derived from these regions must be extracted.

For most achondrite parent bodies, this last bit of indirect evidence is generally the only kind available. The meteorites themselves are the major sources of information on the nature of their parent bodies. When plausible parent bodies are identified, the chemical and the petrologic constraints can be augmented with spectral studies, direct comparison with returned lunar samples, analysis of trapped atmospheric gases, and other methods of inquiry.

Within our collections of achondrites are apparently pieces of the Moon, Mars, main-belt asteroid 4 Vesta, near-Earth asteroid 3103, and a host of other melted asteroids that have not yet been identified. Using these rocks, we have found records of the parent body sizes, the compositions and stratifications of their interiors, and the timing of igneous events. Each achondrite parent body is a natural laboratory, with common igneous processes controlled by variables like composition, temperature, and pressure that differ from body to body. These data would be useful even if we had no idea of where achondrites came from, but their importance is magnified when parent bodies can be identified.

Suggested Readings

There are a number of excellent nontechnical publications on the Moon and Mars, but very little on other achondrite parent bodies. Most of the references cited here are intended for the reader who is interested in more detail, and they may require some additional geologic vocabulary.

GENERAL

Hamblin W. K. and Christiansen E. H. (1990). *Exploring the Planets*, Macmillian, New York.

One of several introductory texts that describe some achondrite parent bodies; Chapters 3 and 5 focus on the geology of the Moon and Mars, respectively, and Chapter 8 briefly introduces differentiated asteroids.

ASTEROID HEATING

Bell J. F., Davis D. R., Hartmann W. K., and Gaffey M. J. (1989). Asteroids: the big picture. In *Asteroids II*, University of Arizona, Tucson, AZ, pp. 921–948.

Spectra from asteroids in various parts of the main belt are interpreted in terms of their thermal and melting histories.

Grimm R. E. and McSween H. Y. Jr. (1993). Heliocentric zoning of the asteroid belt by aluminum-26 heating. Science **259**, 653–655.

This paper quantifies the effects of ^{26}Al decay in bodies formed at different times and locations in the nebula.

ASTEROID PARENT BODIES

Binzel R. P., Gaffey M. J., Thomas P. C., Zellner B. H., Storrs A. D., and Wells E. N. (1997). Geologic mapping of Vesta from 1994 Hubble Space Telescope images. Icarus **128**, 95–103.

An interpretation of the rotational spectra of Vesta in terms of the distribution of HED meteorites on its surface.

Thomas P. C., Binzel R. P., Gaffey M. J., Storrs A. D., Wells E. N., and Zellner B. H. (1997). Impact excavation on asteroid 4 Vesta: Hubble Space Telescope results. Science **277**, 1492–1495.

This paper reports the discovery of a huge crater on Vesta and describes its probable relationship to small Vesta-like asteroids located nearby.

Gaffey M. J., Reed K. L., and Kelley M. S. (1992). Relationship of E-type asteroid 3103 (1982BB) to the enstatite achondrite meteorites and the Hungaria asteroids. Icarus **100**, 95–109.

An analysis of the spectrum and orbit of a near-Earth asteroid and its possible relationship to aubrites.

Goodrich C. A. (1992). Ureilites: a critical review. Meteoritics **27**, 327–352.

This thorough review provides a critical summary of the many published models for the ureilite parent body.

THE MOON

Taylor S. R. (1982). *Planetary Science: A Lunar Perspective*, Lunar and Planetary Institute, Houston, TX.

This marvelous book details what has been learned about lunar geology since the return of Apollo samples.

Heiken G. H., Vaniman D. T., and French B. M., eds. (1991). *Lunar Source Book: A User's Guide to the Moon*, Cambridge U. Press, New York.

More than you ever wanted to know about the Earth's neighbor.

Wilhelms D. E. (1993). *To a Rocky Moon: A Geologist's History of Lunar Exploration*, University of Arizona, Tucson, AZ.

This account of the exploration of the Moon and the lunatics that did it is fascinating and enjoyable.

MARS

Kieffer H. H., Jakosky B. M., Snyder C. W., and Matthews M. S., eds. (1992). *Mars*, University of Arizona, Tucson, AZ.

More than you ever wanted to know about the red planet.

McSween H. Y. Jr. (1994). What we have learned about Mars from SNC meteorites. Meteoritics **29**, 757–779.

A comprehensive review of the geologic history of the red planet, as inferred from studies of SNC achondrites.

6 Iron and Stony–Iron Meteorites

One of the most significant meteorite finds in Antarctica was made accidentally by a New Zealand geologic field party in 1978. While working on the slopes of Derrick Peak, a remote nunatak free of ice and snow, this group stumbled on a number of iron meteorites. News of their discovery was soon transmitted by radio to a joint American–Japanese meteorite search team at a nearby base camp on the Darwin glacier. Team scientists quickly boarded a helicopter and arrived in time to aid in the search for more specimens. In all, sixteen samples were recovered on Derrick Peak. Several, like the meteorite in Figure 6.1, were quite handsome and of sufficient size to make carrying them down the steep mountainside quite difficult.

It seems appropriate that iron meteorites like those at Derrick Peak would be found in Antarctica, the most inaccessible and inhospitable of places. These chunks of metal were originally forged in the deep and once-infernal interiors of extraterrestrial bodies, and it somehow seems right that we endure some hardship to obtain them. This chapter shows that the information carried in iron and stony–iron meteorites is worth all that effort.

The Core of the Problem

Because the Earth's bulk composition is approximately chondritic, its most important element by weight must be iron, comprising nearly 40% of the whole planet. Aside from a few ore deposits that contain iron concentrated by igneous or sedimentary processes, no common rocks contain anywhere near the chondritic abundance of this element. So where is all this missing iron? It is buried in a massive central core of metal, approximately 6,940 km across, that formed during the differentiation of the planet.

Ample evidence for the existence of this core comes from a variety of geophysical measurements. The mean density (mass/volume) of the Earth, 4.0 g/cm^3 after correction has been made for compression of the

Figure 6.1: One of the most spectacular meteorite finds in Antarctica was the Derrick Peak 78009 iron meteorite. Sixteen specimens like this one were found among angular blocks of sandstone on the slopes of the nunatak. The counter in the lower left, used for photographic documentation of samples in the field, is approximately 6 cm long. Photograph courtesy of NASA.

interior due to overlying materials, is significantly greater than that of common rocks. This indicates that more massive material must be hidden in the planet's interior, and iron metal is certainly denser than any kind of rock. Earthquakes produce seismic vibrations with periods of a few seconds that are transmitted through the Earth with velocities dependent on the densities of the matter through which they pass. Seismic waves reflected off the core permit its size to be determined, and the measured velocities of seismic waves propagated through the core allow an estimate of its density and, from that, an inference of its composition. Interior disturbances may also produce gross vibrations of the whole planet acting as if it were a ringing bell. These so-called free oscillations, with periods ranging from minutes to several days, constrain how materials of different densities are distributed inside the Earth. It is well known that the Earth is not perfectly spherical, but bulges at the equator. This flattening, which is due to rotation, is also an expression of the way mass is distributed internally. The moment of inertia, calculated from this rotational flattening, as well as constraints from free oscillations, indicates a large concentration of mass at the planet's center.

The Earth's core is inaccessible to us for direct study, so we must rely almost completely on these indirect measurements. One writer has noted that reaching conclusions from analyses of tremors and jiggles is like trying to reconstruct the inside of a piano from the sounds it makes while crashing down a staircase. Despite the difficulties, geophysical measurements provide a picture of the core that cannot be obtained in any other way. The velocities of seismic waves traveling through the core are best matched by the properties of an iron–nickel alloy, solid at the center and molten on the outside.

Some asteroids must also have formed cores, as evidenced by the existence of iron meteorites. More indirect evidence for asteroidal core formation comes from the study of HED meteorites, which contain little or no metal, although their originally chondritic parent body must have contained appreciable metallic iron. The missing metal is thought to have segregated into a core. There is thus a kind of complementary relationship between achondrites and irons, and it is probably a valid assumption that bodies with differentiated cores also contained achondrites.

Metal-Loving Elements

The idea that other elements might follow iron and nickel into the core was first explored by geochemist Victor Goldschmidt at the University of Oslo in the early part of this century. Goldschmidt's pioneering contributions in understanding the Earth's differentiation were based on meteorites, which he suggested were natural experiments with molten metal, sulfide, and silicate, now frozen into solid form. From chemical analyses of these phases, he determined how various elements partitioned themselves among these substances, and he coined the terms **siderophile**, **chalcophile**, and **lithophile** to describe elements with affinities for metal, sulfide, and rock (either silicate or oxide), respectively. Goldschmidt and his co-workers spent years painstakingly separating and analyzing numerous meteorite components in order to predict their geochemical preferences. The conclusions of this work have been confirmed by studies of industrial smelting products that contain metallic iron, a sulfide matte, and silicate slag. Although Goldschmidt's geochemical affinities have proved to be a serviceable geochemical classification, not all elements can be neatly pigeonholed, because their behaviors can vary with temperature and pressure or can change, depending on what other elements are present.

Core formation, whether within planets or planetesimals, has the effect of scavenging siderophile (metal-loving) elements. Those elements exhibiting the most rabid siderophile tendencies are iron, nickel, cobalt, platinum, osmium, iridium, gold, palladium, ruthenium, rhodium, and tungsten. Of these elements, only iron and nickel have cosmic abundances high enough to make them volumetrically important, so iron meteorites are composed mostly of iron–nickel alloys. The range of nickel in iron meteorites is illustrated in Figure 6.2. No iron meteorites contain less than 5-wt. % nickel, and only a handful contain more than 20 wt. %. Several other elements that exhibit siderophile behavior under the right conditions are germanium, gallium, phosphorus, and carbon. These occur in only minor quantities in iron meteorites, but some are important for classification purposes.

Because siderophile elements are sequestered within metallic cores, achondritic meteorites derived by partial melting of the complementary

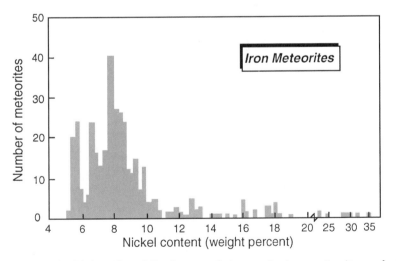

Figure 6.2: Of the siderophile elements that comprise iron meteorites, only iron and nickel have cosmic abundances high enough to occur in major proportions. This diagram summarizes the compositions of hundreds of analyzed iron meteorites. Nickel contents are highly variable, but most metorites contain between 6% and 10% of this element. No iron meteorites contain less than 5% nickel, and only a handful contain more than 20%.

mantles must be depleted in these elements. Consider, for example, how the abundances of tungsten (a siderophile element) and lanthanum (a lithophile element), differ in mantles, as shown schematically in Figure 6.3. During core separation, tungsten is mostly lost from an initially chondritic mantle (vertical arrow), but lanthanum is unaffected. During subsequent partial melting of the mantle to produce basaltic magmas in the absence of metal, both tungsten and lanthanum behave incompatibly and are partitioned together into the melts (diagonal arrow). Fractional crystallization of this melt leads to further concentration of these elements in the liquid, ultimately producing rocks with a range of tungsten-versus-lanthanum abundances that define a diagonal line. The observed depletion of tungsten relative to lanthanum in HED meteorites (see Figure 6.3) shows that its parent asteroid experienced core formation, as did the Moon and Mars.

The compositions of iron meteorites are not restricted only to siderophile elements, although these comprise the great bulk of such meteorites. Experiments have disclosed that a mixture of iron–nickel metal with sulfur has a lower melting point than metal alone. If iron meteorites formed by partial melting, it is to be expected that they should contain some

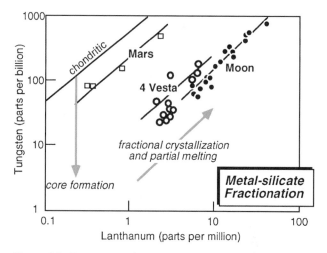

Figure 6.3: During core formation, tungsten (a siderophile element) is partitioned into metal and thus depleted from the mantle, while lanthanum (a lithophile element) is retained in mantle silicates. Subsequent partial melting of the mantle and fractional crystallization of the melt causes both lanthanum and any remaining tungsten to be concentrated together in magmas. Therefore rock samples from differentiated asteroids like the HED parent body are variably depleted in tungsten, falling along sloping lines like those for lunar rocks and Martian meteorites.

sulfur, as indeed they do. Where sulfur goes, chalcophile (sulfide-loving) elements follow. However, iron meteorites contain less sulfur than the amount necessary to make this low-melting-point mixture, suggesting that melting of metal continued after all the sulfide was exhausted.

Crystallization of Irons

Partial melting of chondritic material in asteroids produced liquid metal that, because of its higher density, drained away from residual solids or silicate melt. The commercial smelting of iron depends on this tendency for molten metal and silicate to separate. Liquid in contact with a solid can form either a thin film or beads (see Figure 6.4). Separation of a liquid film is easier because the metallic liquid flows downward through the interconnected spaces at grain boundaries. If the metal instead forms molten beads, they must sink through the silicate mesh, which normally requires that the mesh be disrupted by large degrees of partial melting. Some experimental observations of molten iron in solid silicate suggest that it forms beads and that core formation can occur only when the degree of partial melting of the silicate is large, perhaps more than 50%. However, the apparent loss of metal-sulfide liquid from primitive achondrites (acapulcoites and lodranites) at much lower degrees of partial melting demonstrates that draining of metal did occur efficiently in some asteroidal bodies. Metal-sulfide liquids may move more easily than liquid metal through the enclosing silicate rock.

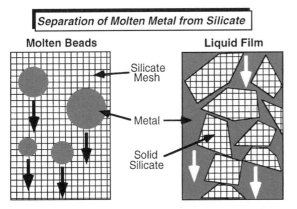

Figure 6.4: These sketches compare the separation of molten iron metal from the enclosing silicate, depending on whether the metallic liquid forms beads or a film between silicate grains. Sinking of molten beads may require that the silicates undergo significant melting to disrupt the mesh.

The crystallization of molten siderophile and, to a lesser extent, chalcophile elements to form iron meteorites is a more complex phenomenon than it might first appear. Over forty different minerals, a number of which are unknown in terrestrial rocks, have been identified in iron meteorites; however, most of these are present in very minor quantities. Only a few of the most important minerals are considered here.

Iron–nickel alloys, which comprise the great bulk of iron meteorites, are of two basic types: Kamacite contains up to 7.5-wt. % nickel, and taenite varies in its nickel content between approximately 20 and 50 wt. %. Although both minerals have crystal structures with cubic symmetry, the different sizes of nickel and iron atoms cause these minerals to have different architectural styles, as illustrated in Figure 6.5. Kamacite forms a body-centered lattice, with each atom located at the center of a cube, so that it is surrounded by eight neighboring atoms. The face-centered lattice of taenite features an atom centered on each face of a cube, so that every atom in this structure has twelve neighbors. It should be obvious that the face-centered lattice is a more efficient way of packing atoms within a given volume. Kamacite and taenite are important components of steel and are known to metallurgists as alpha iron and gamma iron, respectively.

Masses of liquid metal solidified to form iron meteorites as they cooled below roughly 1,400 °C. However, the crystallization histories of these metal chunks did not stop there, but continued as solid-state recrystallization occurred at lower temperatures. Shown in Figure 6.6 is the **phase diagram** for the iron–nickel system below 1,000 °C. Phase diagrams are useful for showing the fields of stability of various minerals in terms of temperature and composition. Above 900 °C, metal of any composition has the taenite structure, but at lower temperatures low-nickel kamacite and high-nickel taenite begin to unmix in the solid state. This unmixing takes place when metal cools to the boundary of the taenite + kamacite field on the phase diagram. From the slope of this boundary, it is apparent that the temperature at which this happens depends on the nickel content of the metal. Kamacite continues to form down to approximately 500 °C, at which point migration of atoms through the solid metal becomes so sluggish that unmixing stops.

To see how iron meteorites of different compositions recrystallize, let us take three specific examples, illustrated by arrows on the iron–nickel phase diagram (Figure 6.6). These examples are simplified, because the iron–nickel phase diagram is very complex when viewed in detail. On cooling, an alloy containing 5% nickel will begin to change from the taenite structure to the kamacite structure at approximately 800 °C. Below

Crystal Structures of Metals

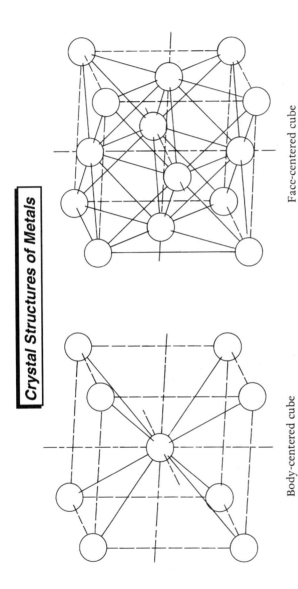

Body-centered cube

Face-centered cube

Figure 6.5: Iron–nickel alloys, from which iron meteorites are formed, exhibit two basic crystal structures. Kamacite forms a body-centered lattice, with each atom located at the center of a cube so that it is surrounded by eight other atoms. Taenite occurs as face-centered cubes, with an atom centered on each face. In this structure, each atom is surrounded by twelve neighboring atoms.

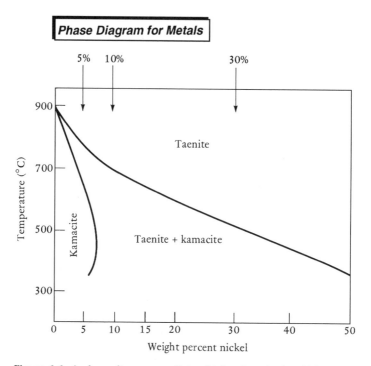

Figure 6.6: A phase diagram predicts which minerals should be stable under various combinations of temperature and chemical composition. This simplified phase diagram for the iron–nickel system below 1,000 °C, constructed from a series of experiments by metallugists, is divided into three fields showing the stability of taenite, kamacite, or both. Cooling of an iron meteorite with a certain nickel composition would be represented on this diagram by the dropping of a vertical line from the top. A mass of solid metal will change its structure as it moves into a different stability field during cooling.

approximately 650 °C this metal will enter the kamacite field, and the transformation will be complete. The final product will consist only of kamacite. At the other end of the compositional scale, a cooling alloy with 30% nickel will just about reach the temperature limit for atom migration (500 °C) before any unmixing starts. This meteorite, when finally cooled, will consist almost entirely of taenite. The most interesting situation is something in between these two extremes, as illustrated by a 10% nickel alloy. It also turns out that intermediate nickel values are the most common in iron meteorite compositions. In this case, the taenite cools to approximately 700 °C and begins to unmix. By 500 °C, this composition resides within the field where taenite and kamacite coexist, so the final meteorite will contain both minerals.

The physical appearance of iron meteorites containing both kamacite and taenite is striking. Grains of kamacite grow in certain preferred orientations within the face-centered taenite lattice. Kamacite forms four sets of plates that cut the corners of the original cubic crystals, forming octahedra. The intergrown plates of kamacite within taenite can be readily seen in slices of iron meteorites that have been polished and etched with dilute acid. In 1808, Count Aloys de Widmanstätten, director of the Imperial Porcelain Works in Vienna, first observed this structure, now called the **Widmanstätten pattern**. The geometry of the intergrown kamacite plates, of course, varies depending on the orientation of the meteorite when it was sawed. Sketches of how the Widmanstätten pattern would apear in various sections through a parent taenite crystal are shown in Figure 6.7.

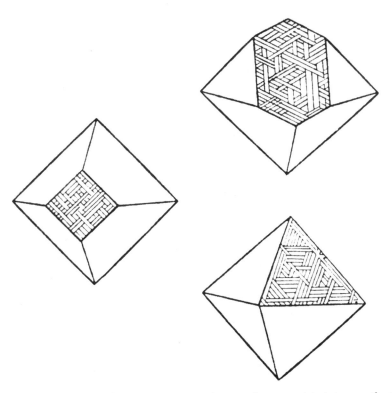

Figure 6.7: The Widmanstätten pattern is a regular geometric intergrowth of kamacite and taenite that forms during slow cooling of iron meteorites with appropriate composition. These sketches of the Widmanstätten pattern as it would appear in various cuts through the original taenite crystal were made by mineralogist Gustav Tschermak in 1894.

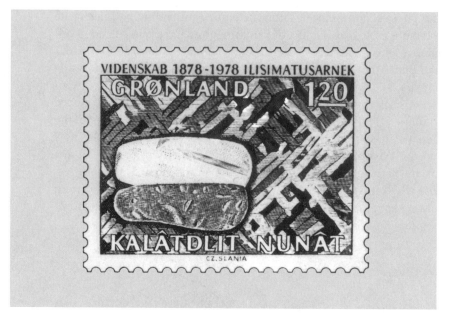

Figure 6.8: This postage stamp commemorates the hundredth anniversary of the discovery of the Cape York iron meteorite in Greenland. It clearly shows the Widmanstätten pattern, with an elongated rounded grain of troilite and the socket from which it came.

Randomly interspersed among the metal grains in most iron meteorites are other minerals also rich in iron or nickel. Troilite (iron sulfide) is almost ubiquitous, forming dark, rounded nodules (see Figure 6.8). Inclusions of this phase were first observed by the Jesuit priest Domenico Troili in 1766, but a century elapsed before accurate chemical analyses proved that it was distinct from pyrite (iron disulfide), the fool's gold commonly found in terrestrial rocks. Sulfur is highly soluble in molten iron but practically insoluble in solid iron, so it is necessary for this element to form a sulfide as the metal crystallizes.

Carbon and phosphorus are also soluble in metallic liquids, and small amounts of these elements (generally less than 0.1%) can be substituted in iron alloys. But, as with sulfur, the bulk of carbon and phosphorus end up in their own minerals. Carbon forms cohenite, a carbide of iron, nickel, and cobalt. This mineral is unstable at all temperatures under conditions of very low pressure and breaks down into metal (kamacite) plus pure carbon (graphite). That cohenite occurs at all is due to the fact that its decomposition is extremely slow, and different meteorites show this breakdown in various stages of arrest. Phosphorus forms the mineral schreibersite (iron–nickel phosphide). This is a particularly attractive mineral, commonly occurring as tiny white prisms within metal grains. Any chromium in the original melt forms chromite (chromium–iron oxide). Most of the other elements in iron meteorites are present in only trace amounts and display enough siderophile behavior to dissolve in metal.

Order Out of Chaos

On casual observation, iron meteorites seem to be highly diverse and perplexing samples, but order is restored by proper classification. A good classification system should do more than simply superficially lump together similar meteorites. Ideally it should provide information on their origins. There are two entirely different classification schemes for iron meteorites now in use. One of these is based on the structures present in irons, and classification can be done by visual inspection. The other requires accurate and time-consuming chemical analyses. Both systems produce similar groupings in most cases, but the chemical classification provides more genetic constraints. We first consider the **structural classification**.

Irons with less than approximately 6-wt. % nickel contain kamacite but no taenite. The sizes of the original kamacite grains must have been larger than the present dimensions of the meteorites, for in most cases these irons are single crystals. Because kamacite is cubic, such meteorites are called **hexahedrites** (abbreviated H), from the Greek word for six-sided. Polished surfaces of these irons are featureless, except for numerous striations in some meteorites. An example of a hexahedrite with such striations, called **Neumann lines**, is shown in Figure 6.9. These lines are boundaries between twins (segments of the same crystal in different orientations) formed by shock deformation.

Octahedrites range in nickel content between approximately 6 and 17 wt. %. These meteorites contain both kamacite and taenite in the decorative Widmanstätten pattern. The octahedrites constitute by far the largest and most diverse group of irons, so it is advantageous to subdivide them. A convenient way of doing this is by making use of the widths of the kamacite bands that form the Widmanstätten pattern. It is conventional to use a fivefold subdivision of octahedrite bandwidths, each formed from the previous smaller one by multiplication times 2.5. In this system, octahedrites are separated into the following groups (with abbreviations in parentheses): coarsest (Ogg), coarse (Og), medium (Om), fine (Of), and finest (Off). Examples of some structural differences among octahedrites are shown in Figures 6.10 and 6.11.

Figure 6.9: The Calico rock (Arkansas) hexahedrite consists almost entirely of kamacite. This cut face shows Neumann lines formed by shock deformation, but is otherwise featureless. The meteorite measures approximately 8 cm across. Photograph courtesy of the Smithsonian Institution.

The finest octahedrites actually are graded into meteorites with no obvious structure. These irons, which have high nickel contents, are called **ataxites** (abbreviated D). The name is derived from the Greek word *ataxia*, meaning without order. This is perhaps an unfortunate term, because ataxites do have a microscopic Widmanstätten pattern, but it is not perceptible to the naked eye. Such meteorites consist almost entirely of taenite, with only a few microscopic plates of kamacite. An ataxite is shown in Figure 6.12.

A number of iron meteorites cannot be conveniently assigned to any of these structural classes. These are commonly rather fine grained, consisting of centimeter-sized crystals rather than the usual larger grains ranging

Figure 6.10: Octahedrites can be classified by the widths of their kamacite plates. This etched slab of the Mount Stirling (Australia) coarse octahedrite exhibits a coarse Widmanstaettian pattern. The large black ovals are inclusions of the sulfide mineral troilite. The vertical cut surface of the meteorite measures approximately 3 cm. Photograph courtesy of the Smithsonian Institution.

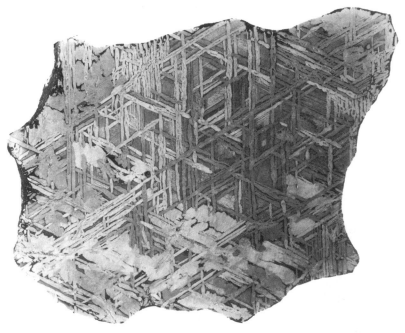

Figure 6.11: The Waingaromia (New Zealand) fine octahedrite contains thin, oriented bands of kamacite that form its Widmanstaettian pattern. The width of this slab is approximately 11 cm. Photograph courtesy of the Smithsonian Institution.

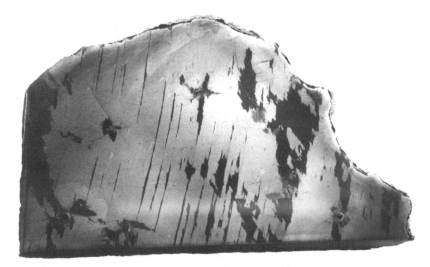

Figure 6.12: Ataxites have no readily observable Widmanstätten structure, because they contain only minute quantitites of kamacite. Hoba (Namibia), the largest recovered meteorite, is of this type. This slab of the Hoba ataxite is featureless except for the dark bands that were produced by shock. The bottom cut face measures approximately 15 cm. Photograph courtesy of the Smithsonian Institution.

up to meter sized. Iron meteorites that resist classification are quite logically called **anomalous irons** (abbreviated Anom).

The chemical classification system for irons utilizes trace elements to diagnose the different **chemical groups**. An intially proposed subdivision into four groups, identified by Roman numerals I through IV, was based on varying contents of gallium and germanium as well as the abundance of nickel. When enough high-quality chemical data became available, some of these groups were subdivided even further. For example, group IV has now been separated into groups IVA and IVB, which apparently have nothing to do with each other. To make things even more complicated, a few subdivided groups have now been recombined, after transitional members were discovered. This has led to the somewhat confusing situation of having some group names like IIAB. The resulting motley array of symbols is very difficult to remember, but otherwise this is a very serviceable classification system. Most of the approximately 600 known iron meteorites have now been analyzed. From this large database John Wasson at the University of California, Los Angeles, has recognized thirteen important groups with five or more members each. When the measured concentrations of siderophile trace elements like germanium and gallium are plotted against nickel on logarithmic scales, these subdivisions

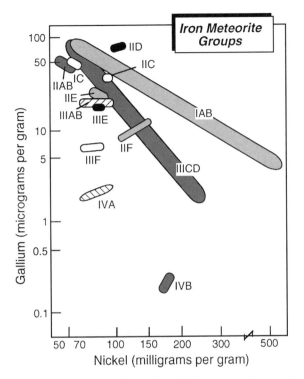

Figure 6.13: A logarithmic plot of the measured concentrations of gallium versus nickel in iron meteorites provides a way to classify them into different chemical groups. The twelve groups shown account for 86% of all analyzed irons, the others being anomalous. Other elements besides gallium and nickel can also be used to classify these meteorites. The meteorites were analyzed by John Wasson and his colleagues (University of California at Los Angeles).

are clearly resolved. Figure 6.13 illustrates the clusters representing these groups on a gallium-versus-nickel abundance plot. Meteorites that do not plot within these well-defined chemical clusters are again called anomalous. Keep in mind that anomalous irons in the chemical sense may not be anomalous in the structural sense.

The meteorites in each chemical group generally belong to a limited range of structural classes, as shown in Table 6.1. This is actually somewhat surprising, because the structural classification is not solely dependent on composition. The kamacite bandwidths on which this system is based are dictated not only by the nickel content of the cooling taenite crystals but also by the rate at which these crystals cooled. The metallurgist who proposed the structural classification for iron meteorites actually cheated a little bit in defining the bandwidth intervals so that they would

Table 6.1: Comparison of Chemical and Structural Classisfications for Iron Meteorites

Chemical Group	Structural Class: H	Ogg	Og	Om	Of	Off	D	Anom	Proportion (%)
IAB			—	—					18.7
IC			—					—	2.1
IIAB	—	—							10.8
IIC						—			1.4
IID				—	—				2.7
IIE			—	—					2.5
IIIAB			—	—					32.3
IIICD					—	—	—		2.4
IIIE			—						1.7
IIIF			—	—	—				1.0
IVA					—				8.3
IVB								—	2.3
Anom	—	—	—	—	—	—	—	—	13.8

most nearly coincide with chemical groups. However, this definition of structural classes is fully justified by the fact that there is a correlation in most cases between band width and chemical grouping. The few iron meteorites for which the correlation does not hold can be rationalized as the products of cooling rates different from those for the rest of their chemical group.

Solidification of Cores

Chondrites, which have never been melted, contain metal that is much more uniform in composition than metal in iron meteorites. For example, the total range of iridium abundances in several groups of irons varies a thousandfold, whereas iridium concentrations in the metal of any particular chondrite class vary over less than a factor of two. These large variations within iron meteorite groups are attributable to the crystallization processes by which cores were formed. Fortunately these variations are not random. Within a given chemical group, irons can be ranked in a sequence such that some elements (nickel, gold, arsenic, cobalt, palladium, phosphorus, and molybdenum) increase while others (iridium, osmium, platinum, ruthenium, tungsten, and chromium) decrease. When plotted on logarithmic scales, the concentrations of such elements in meteorites of any one classified group form nearly straight lines. One example of such trends is shown in Figure 6.14. This consistent pattern from core to core is not accidental, but arose through fractional crystallization as solid metal

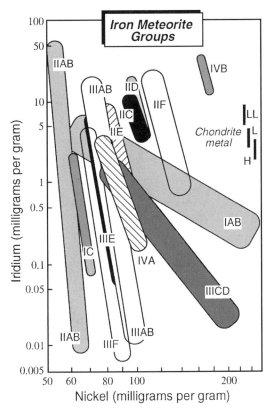

Figure 6.14: This logarithmic plot shows iridium-versus-nickel contents in iron meteorite groups. The iridium abundances extend over many orders of magnitude because of fractional crystallization occurring during solidification of parent body cores.
These large ranges of iridium abundances are very different from the unfractionated
compositions of metals in H, L, and LL ordinary chondrites.

separated from liquid metal. The partitioning of these elements between
solid and liquid metals has been measured in the laboratory by metallurgist Joe Goldstein, formerly of Lehigh University, and these experiments
correctly predict the trends of these elements observed within iron groups.
Nickel and other incompatible elements are concentrated preferentially
in liquid, whereas compatible elements like iridium accumulate in solid
metal. Fractional crystallization that efficiently separates liquid from solid
produces linear trends with negative slopes on the iridium-versus-nickel
diagram shown in Figure 6.14.

Study of the huge Cape York (Greenland) meteorite, a member of the
largest group of irons (IIIAB), suggests that the explanation above is
essentially correct in general but flawed in detail. The average iridium and
nickel concentrations for different samples of this meteorite plot within

the trend for IIIAB irons shown in Figure 6.14, but define their own slope, which is distinctly shallower than this steep trend. Henning Haack and Edward Scott of the University of Hawaii attributed this oddity to solidification of the core as large blades or needles (**dendrites**) growing inward from the core–mantle boundary. As the dendrites enlarged, they isolated pockets of liquid, creating a compositionally heterogeneous core on a small scale. Thus dendritic growth may have produced localized differences in metal composition that caused the IIIAB trend in Figure 6.14 to be wider than if the liquid had been homogenized.

Two groups of iron meteorites, IAB and IIICD, are unusual in that their trends are distinct from those of other groups. In Figures 6.13 and 6.14, the slopes of these two trends are much shallower than those of the other groups. Such trends appear to violate the rules for the igneous behavior of these elements determined from experiments. One possible explanation for this observation is that these two iron groups did not crystallize from complete liquids. Instead, we can model their trace-element patterns successfully by assuming that they represent mixtures of metallic liquids with unmelted iron. Perhaps they formed as pools of incompletely melted metal produced during impacts into a chondritic asteroid. Alternatively, they may represent samples of cores that were particularly rich in sulfur. The presence of sulfur modifies the partitioning behavior of elements like gallium and iridium, resulting in less steep-fractionation trends for these elements relative to nickel.

The effect of fractional crystallization or possibly incomplete melting has been to smear the chemical compositions of individual iron meteorites belonging to any particular group. Although this complexity makes their classification more difficult, it is actually an advantage because it allows us to examine intermediate steps in the partial melting and crystallization processes that affected asteroidal cores.

Irons in the Fire

The reason that the structures of iron meteorites vary with cooling rate is that the Widmanstätten patterns form by unmixing in the solid state, which is a temperature-controlled process. The longer iron meteorites are held at high temperatures in cores within parent bodies, the greater the amount of unmixing. Quantitative measurements of cooling rates constrain the depth of burial of cores and thus specify the minimum sizes of the parent bodies in which the irons formed. How have cooling rates been recorded in these meteorites, and is it possible to play back this record?

To understand this, we must return to the iron–nickel phase diagram. Again consider a mass of metal containing 10% nickel, which will crystallize from liquid into taenite and cool along a vertical line in Figure 6.6. At approximately 700 °C, this alloy will begin unmixing to form oriented bands of kamacite, the composition of which is approximately 4% nickel. The boundaries to the taenite + kamacite field give the compositions of the two metal phases coexisting at any one temperature. These sloping field boundaries indicate that the nickel contents of both kamacite and taenite increase with further cooling. For example, when the mass has cooled to 600 °C, the kamacite composition is approximately 6% nickel and that of taenite is approximately 19% nickel. How can this be? The only way that both metal phases can simultaneously increase their nickel contents while the composition of the whole mass remains constant is to increase the amount of the low-nickel phase (kamacite) at the expense of the high-nickel phase (taenite). The kamacite plates must become thicker, and the taenite between them must shrink by an equivalent amount.

The growth or shrinking of metal phases occurs by migration of atoms within the solid metal grains. Nickel atoms at a grain margin creep into the interior by successive exchanges with iron atoms until the whole grain is homogeneous in composition. This process, called **diffusion**, occurs more rapidly at high temperatures than at lower ones. At some point, approximately 700 °C for taenite, diffusion becomes sluggish enough that nickel atoms can no longer travel freely into grain interiors, and they begin to pile up at the margins. This situation is like a traffic jam, in which constriction of the flow of cars occurs first at one point and then propagates back along the highway. This slowdown of diffusion results in chemically zoned taenite in which the nickel content decreases toward the center of the grains. By approximately 500 °C, diffusion effectively grinds to a halt, and the zoning profile is frozen in. An example is shown in Figure 6.15. Chemical analyses of microscopic spots in a traverse through a taenite grain from one side to the other define an M-shaped nickel profile, with high nickel at the edges and lower nickel in the center.

Widmanstätten patterns formed at cooling rates so slow that they cannot be easily reproduced in the laboratory. Nevertheless, understanding how these patterns formed permits the growth of this pattern to be modeled theoretically. The final nickel distribution across any taenite crystal in an intergrowth with kamacite will depend on the bulk nickel content of the mass and the size of the grains through which diffusion occurred. These can be measured in the laboratory, along with the diffusion rates

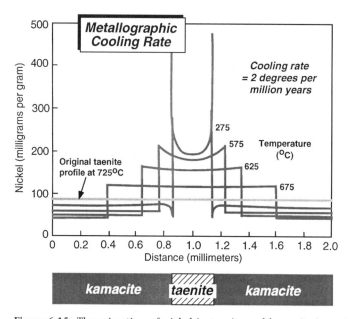

Figure 6.15: The migration of nickel in taenite and kamacite in an iron meteorite cooling at 2° per million years is simulated at several temperatures in this diagram. Initially the meteorite consists entirely of taenite. At 675 °C, kamacite begins to form, and the kamacite bands become progressively thicker at the expense of taenite as the temperature is lowered further. The nickel content in taenite decreases away from the margins because diffusion during cooling becomes too sluggish at low temperatures to homogenize the crystal. The resulting M-shaped nickel profile in taenite provides a means of calculating the cooling rates of iron meteorites. The analysis of nickel in meteoritic taenite and the distance from the analysis point to the nearest kamacite boundary can be compared with calculations of the profiles at various cooling rates. This numerical simulation was made by Kaare Rasmussen (University of Odense, Denmark).

of nickel at various temperatures. John Wood of the Smithsonian Astrophysical Observatory first used these data to develop computer models for the growth of kamacite bands and to calculate the resulting nickel zoning profiles in taenite grains at different cooling rates. The only action that is then required for determining the cooling rates at which iron meteorites formed is to measure the M-shaped profiles in actual taenite grains and compare these with the theoretical profiles.

Cooling rates derived from this method vary from less than one degree to several thousand degrees per million years. These rates may seem slow, yet they are still appropriate to small asteroidal bodies. The cores of large planetary bodies like the Earth have not yet cooled to the point where they are completely solid in nearly 4.5 billion years.

Silicate Inclusions in Irons

The scarcity of silicates in iron meteorites is not surprising, because the radical differences in density between these materials presumably caused most of the lighter silicates to float out of metallic cores. However, several classes of irons do contain **silicate inclusions**, and in some cases they are reasonably abundant (see Figure 6.16).

We have already seen that oxygen isotopes provide a valuable tool for indicating possible relationships between stony meteorites. Oxygen isotopes have also been measured in silicate inclusions, and the data are summarized in Figure 6.17. Silicates in IIIAB irons are similar in composition to HED achondrites. However, because the HED parent body (4 Vesta) is thought to be still intact, it is unlikely that its core has been exposed and sampled. This similarity probably demonstrates that more than one asteroid can share the same oxygen isotopic signature (as do the Earth, Moon, and aubrite parent asteroid).

Inclusions in IIE irons define a mass-fractionation line through H chondrites, and the mineralogy of some of them resembles that of H chondrites.

Figure 6.16: Some iron meteorites, like the Pitts (Georgia) octahedrite, contain silicate inclusions. The large gray areas with irregular shapes in this photograph are the inclusions, each of which is 1–2 cm in diameter. Silicate inclusions are useful sources of information that cannot be obtained from metallic minerals. Photograph courtesy of the Smithsonian Institution.

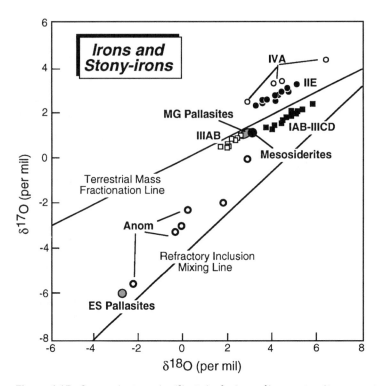

Figure 6.17: Oxygen isotopes in silicate inclusions of iron meteorites suggest possible affinities with other groups of chondrites and achondrites. Inclusions in the IIE and IVA iron groups are similar to those of H and L ordinary chondrites, respectively. Inclusions in IAB and IIICD irons overlap those of winonaites (a group of primitive achondrites), whereas those in IIIAB irons resemble HED achondrites. The fields for silicates in several types of stony–irons (main-group MG pallasites and mesosiderites) also plot along the HED mass-fractionation line. Other pallasites, such as the Eagle Station ES group and anomalous irons plot in different fields. All meteorites were analyzed by Robert Clayton and colleagues (University of Chicago).

Group IVA inclusions plot along a similar line through L and LL chondrites, but their mineralogy (consisting only of pyroxene and silica) is clearly different from that of chondrites. These iron meteorite groups may represent the cores of differentiated objects that were formerly ordinary chondrite parent bodies, or they may contain unrelated silicate materials added to molten metal during impacts. Several anomalous irons contain inclusions with oxygen isotopes that plot near carbonaceous chondrites.

Groups IAB and IIICD irons have been suggested to be incompletely melted samples. The oxygen isotopic compositions of their silicate inclusions are different from all known chondrite groups, but are similar to those of the **winonaites** (a group of primitive achondrites). The minèral compositions in the silicate inclusions are also similar to winonaites. In this case, it seems likely that all these meteorites might have been derived from the same asteroid.

As Old as Iron

The crystallization ages of iron meteorites have been measured with a chronometer based on a radiogenic isotope of osmium, ^{187}Os, which forms by decay of a radioactive isotope of rhenium, ^{187}Re. Both elements are siderophile and are therefore concentrated in metal. Meteorites from groups IAB, IIAB, IIIAB, IVA, and IVB all plot along the same isochron, corresponding to an age of 4.61 ± 0.01 billion years. There is some uncertainty as to the rate at which ^{187}Re decays, so this ancient age may be slightly too old. It is clear, however, that the parent asteroids for iron meteorites differentiated very early, at essentially the same time that most achondritic meteorites formed.

The nonmetallic silicate inclusions in irons can also be radiometrically dated. Rubidium and strontium isotopes indicate that group IAB silicate inclusions are 4.45 ± 0.01 billion years old, just a few million years younger than the ages of chondrites. These particular inclusions appear to be recrystallized primitive achondrites (winonaites), and their ages presumably record the time of partial melting. The rounded silicate inclusions in IIE irons have ages of 4.6 ± 0.1 billion years, except for inclusions in one meteorite that give an age of approximately 3.8 billion years. These ages may record the time of immersion of chondritic material in molten silicate. If this last age is correct, it could suggest that differentiation of the IIE parent body was a protracted process.

Better age resolution is provided by the presence of extinct radionuclides in irons. Hafnium and tungsten are both highly refractory elements that were apparently present in chondritic proportions in meteorite parent bodies. Hafnium is lithophile and tungsten is siderophile, leading to separation of these elements as tungsten is partitioned into metal during core formation. Radioactive ^{182}Hf has a half-life of only 9 million years, so if core formation occurred while this short-lived isotope was still alive, hafnium and later its tungsten decay product, ^{182}W, would be concentrated in the silicate mantle and depleted in the core (relative to chondrites, which experienced no metal-silicate fractionation). Measurements by Der-Chuen Lee and Alex Halliday of the University of Michigan reveal that IAB and IIAB irons are depleted in ^{182}W by an amount corresponding to their formation within the first 15 million years of the solar system's history. Likewise, measurements of excess amounts of an isotope of silver, ^{107}Ag, formed from rapid decay of ^{107}Pd (a short-lived isotope of the siderophile element palladium), by J. Chen and Jerry Wasserburg at Caltech, indicate that this extinct radionuclide was present in the iron groups IIAB, IIIAB, IVA, IVB, and several anomalous irons. These data are consistent with a narrow time interval of only 12 million years for formation of all these irons.

Iron meteorites are very old, similar in age to that of some groups of achondrites and chondrites. The same heat source(s) that caused metamorphism and silicate melting must have produced molten metal that solidified to form cores.

Added Complications

Iron meteorites are already difficult enough to understand, but nature and man in some cases have joined forces to make this task even more taxing. These added complications take the form of new structural modifications superimposed on the original ones.

One cause of these changes is deformation due to shock, either by impacts in space or the sudden deceleration on arrival on the Earth. Iron meteorites are not commonly brecciated as are stony meteorites, but the consequences of impacts are nevertheless apparent. One prominent example is the development of Neumann lines, due to twinning, already illustrated in Figure 6.9.

Another kind of change has been induced by heating, which anneals the original structures. Friction during atmospheric passage produces heat, but usually an annealed exterior rind of only a few millimeters thickness is created by this rapid transit. However, of the 600 or so known irons, 94 have been heated artificially by man in his curiosity and eagerness to test or utilize these metals. A number of iron meteorites have also been used as raw materials to make various artifacts, such as decorative beads or swords (see Figure 6.18). Furthermore, our forefathers showed a propensity to maim iron meteorites by hammering and chiseling them. The Widmanstätten patterns of eight large irons were distorted by such cold working when these masses were used as anvils.

Pallasites

From the name **pallasite**, one might logically (but incorrectly) infer that these are samples of the asteroid 2 Pallas. Actually, the only link between the asteroid and the meteorites of this class is that they take their names

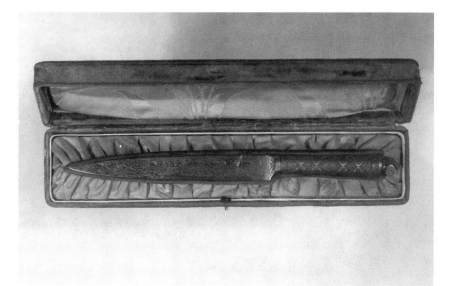

Figure 6.18: Because they are composed of metal, iron meteorites have been used in the past to make artifacts. This Jalandahr knife was forged from an iron meteorite that fell in Punjab (India) in 1621. Photograph courtesy of the Smithsonian Institution.

from the same person, the German naturalist Pyotr Pallas. In 1776, Pallas wrote a lengthy description of one of these meteorites. He had earlier been invited by the Russian monarch, Catherine the Great, to explore the vast uncharted region of Siberia, and one of the curiosities he obtained on the expedition was the first known pallasite.

Pallasites are stony–iron meteorites composed almost exclusively of abundant olivine crystals enclosed in metal (see Figure 6.19). These are particularly attractive meteorites, and the green olivine crystals are often of gem quality (olivine gemstones are known as peridots). The relative amounts of these two constituents vary, but an olivine-to-metal volume ratio of approximately 2 to 1 is common. This is approximately the predicted proportion of olivine where the silicate grains are close-packed spheres of uniform size. Where metal is more abundant, it has a well-developed Widmanstaettian pattern.

The composition of olivine in most pallasites is very rich in magnesium, and metals have nickel, germanium, and gallium contents that are rather similar to those in group IIIAB irons (the most populous iron group, consisting of approximately 200 meteorites). Most pallasite metals actually define an extension of the IIIAB trend. The composition of pallasite metal appears to be that of the liquid that would be left after extensive crystallization of a IIIAB melt.

Figure 6.19: Pallasites are stony–iron meteorites consisting of green olivine crystals enclosed by metal. This specimen from Salta (Argentina) measures approximately 14 cm in its short dimension. Photograph courtesy of the Smithsonian Institution.

A few pallasites, typified by the Eagle Station (Kentucky) meteorite, contain iron-rich olivines and metals with compositions unlike those of other iron groups. It is therefore useful to distinguish this Eagle Station group from main group pallasites. These groups also differ markedly in the oxygen isotopic compositions measured for their olivines, as shown in Figure 6.17. The two groups must have had origins in different parent bodies, although they look superficially similar.

Mesosiderites

Mesosiderites are stony–iron meteorites that have sometimes been called "messysiderites" because they are agglomerations of such different materials (see Figure 6.20). The thirty or so mesosiderites consist of roughly equal proportions of metal and silicates. In fact, the roots of the name are the Greek words *mesos* for middle and *sideros* for iron. The chief problem in understanding these meteorites is that these two components apparently had nothing to do with each other – they are just accidental mixtures.

The silicate fraction of mesosiderites consists mostly of pyroxenes and plagioclase. Many of the silicate fragments in these meteorites are very similar to eucrites, diogenites, and howardites, basically igneous rocks

Figure 6.20: Mesosiderites are an important class of stony–iron meteorites. The Reckling Peak A79015 (Antarctica) mesosiderite contains dark gray silicates embedded in more highly reflective metal. The cube measures 1 cm on a side. Photograph courtesy of NASA.

from the crust of an achondrite parent body. What is noticeably missing from mesosiderites is olivine, which is the most abundant mineral in chondrites and should occur in large quantities in the mantles of any differentiated asteroids. Thus mesosiderites can be viewed as mixtures of core and crustal materials, but without samples of the intervening mantle that would have separated these components in any plausible parent body. Most of the rocky materials have been strongly recrystallized and metamorphosed, probably as a result of their immersion in molten iron.

Metals in mesosiderites show very limited variations in their trace-element abundances, unlike the wide chemical ranges for irons already seen in previous figures. They are most similar to the metals in IIIAB irons, but without the strong fractionation trends characteristic of that core. For example, the ratio of iridium to nickel in mesosiderites varies by less than a factor of two, whereas the same ratio in IIIAB irons has a range of several thousand. Thus a mixture of metal and silicate must have occurred early in the core's history, before the metallic liquid had a chance to fractionate.

The texture of mesosiderites indicates that the metal phase was molten at the time it was mixed with silicate. An obvious way to accomplish mixing is by a collision between two already differentiated asteroids, allowing the still-liquid core of one body to mix with the solidified crust of the other. The absence of mesosiderite analogs, even from craters known to have been formed by impacting irons (such as Meteor Crater, Arizona), may result from the high velocities that characterize such events on the Earth and the Moon. In the early solar system, collisions between asteroids may have been less energetic, so that mixing of projectiles was more common.

A basaltic clast in the Vaca Muerta (Chile) mesosiderite gives a radiometric age of 4.48 ± 0.09 billion years, roughly the same as for other iron and achondritic meteorites. Somehow, before that time, a collision must have mixed these disparate materials. The resulting mixture was buried deeply, so that slow cooling produced a Widmanstätten pattern in the metal. The cooling rates for mesosiderites measured from nickel profiles in taenite are exceptionally slow, less than half a degree per million years. No satisfactory explanation for how these meteorites could have cooled so slowly has been offered, unless they formed within a very large asteroid.

Precious Metals

The massive core of the Earth is now and always will be beyond our scientific grasp, at least in terms of direct analysis of samples. However, among the more amazing gifts that nature bestows are occasional samples of the cores of other solar system bodies. Iron and stony–iron meteorites carry a record of differentiation events probably similar to those that affected the interior of our own world.

The properties of iron meteorites indicate that they originally formed as segregations of molten metal deep inside asteroids. The separation of metal from silicates was largely complete in most cases, and the chemical trends observed in most irons match those predicted by laboratory crystallization experiments. The few exceptions either formed by incomplete melting of metal or by fractional crystallization in sulfur-rich cores. Cooling was slow enough to produce Widmanstätten patterns in irons that had appropriate concentrations of nickel. Oxygen isotopes in sparse silicate inclusions in a few iron groups suggest possible affinities with chondrite and achondrite parent bodies. Radiometric ages of the irons and their inclusions indicate that differentiation happened early in the history of these bodies and probably accompanied the partial melting of silicate mantles to produce achondrite magmas.

Studying the core-forming process as recorded in iron meteorites has been the focus of this chapter. In Chapter 7 we try to reconstruct some of the parent bodies for irons and stony–irons. Identification of possible examples of these bodies may also provide insights into core formation.

Suggested Readings

The first reference provides an excellent introduction to iron meteorites and is profusely illustrated. Most of the other works cited tend to be demanding, but they contain a great deal of detailed information and interpretation of these objects.

GENERAL

Buchwald V. F. (1975). *Handbook of Iron Meteorites*, University of California, Berkeley, CA.

An authoritative and exhaustive technical reference that provides a lucid overview of the field and detailed individual descriptions for most iron meteorites.

Dodd R. T. (1986). *Thunderstones and Shooting Stars: The Meaning of Meteorites*, Harvard U. Press, Cambridge, MA.

Chapter 8 of this easy-to-digest book gives a wonderful summary of iron meteorites and pallasites.

CLASSIFICATION OF IRONS

Scott E. R. D. and Wasson J. T. (1975). Classification and properties of iron meteorites. Rev. of Geophys. Space Phys. **13**, 527–546.

A technical review focusing on the chemical classification of irons.

Wasson J. T. (1974). *Meteorites*, Springer-Verlag, Berlin.

A technical monograph containing useful classification tables for most known iron meteorites.

MINERALOGY OF IRONS

Buchwald V. F. (1977). The mineralogy of iron meteorites. Philos. Trans. R. Soc. London A **286**, 453–491.

A technical paper summarizing the many minerals that have been found in iron meteorites.

THERMAL HISTORIES OF IRONS

Wood J. A. (1964). The cooling rates and parent planets of several iron meteorites. Icarus **3**, 429–459.

The classic paper that first outlined the principle of metallographic cooling rates.

Saikumar V. and Goldstein J. I. (1988). An evaluation of the methods to determine the cooling rates of iron meteorites. Geochim. Cosmochim. Acta **52**, 715–726.

This paper revises metallographic cooling rates to be five times faster than those determined by Wood above, based on a new iron–nickel phase diagram and diffusion rates of nickel.

CHRONOLOGY OF IRONS

Shen J. J., Papanastassiou D. A., and Wasserburg G. J. (1996). Precise Re–Os determinations and systematics of iron meteorites. Geochim. Cosmochim. Acta **60**, 2887–2900.

Measurement of the radiometric ages of various iron groups by use of isotopes of siderophile elements.

Lee D.-C. and Halliday A. N. (1995). Hafnium–tungsten chronometry and the timing of terrestrial core formation. Nature (London) **378**, 771–774.

Application of a now-extinct radionuclide to determine the time of asteroid core formation.

STONY–IRONS

Buseck P. R. (1977). Pallasite meteorites – mineralogy, petrology and geochemistry. Geochim. Cosmochim. Acta **41**, 711–740.

A technical paper describing the properties of pallasites.

Rubin A. E. (1997). A history of the mesosiderite asteroid. Am. Sci. **85**, 26–35.

An easily understood, popularized article summarizing the properties and origin of mesosiderites.

Haack H., Scott E. R. D., and Rasmussen K. L. (1996). Thermal and shock history of mesosiderites and their large parent asteroid. Geochim. Cosmochim. Acta **60**, 2609–2619.

This technical paper presents an alternative to the origin for mesosiderites envisioned by Rubin.

7 Iron and Stony–Iron Parent Bodies

he relative emptiness of space between 2.0 and 3.5 AU has been anything but a sanctuary for preservation of the smaller bodies of the solar system. The planetesimals that comprise the present asteroid belt are mostly relics of some terrible destruction, a cosmic Armageddon that has left only a few bodies intact (see Figure 7.1). Nowhere is this seen more clearly than in the parent bodies of irons and stony–irons. These meteorites are samples of the metallic skeletons that once must have undergirded the rocky parts of differentiated asteroids. Their parent bodies, if they still exist, are denuded cores from which the overlying silicate mantles and crusts have been stripped off and reduced to rubble.

In this chapter we attempt to determine the characteristics that might allow us to recognize orbiting metallic cores. These properties are then used to identify a few possible examples from among the debris littering the interplanetary battleground otherwise known as the asteroid belt.

Core Sizes

The proportions of core materials in differentiated asteroids probably approximate the relative amounts of metal (plus sulfide) versus silicates in chondrites, the presumed precursors of differentiated asteroids. Chondrites commonly contain 8%–20% metal and up to 15% sulfide by volume, so as much as a third of the volume of a differentiated asteroid might plausibly be core material.

The cooling rates determined from nickel diffusion in the metal of iron and stony–iron meteorites, summarized for most groups in Figure 7.2, require that they were once buried within objects no larger than respectable asteroids. The vertical bars on the right-hand side of this diagram show depth-of-burial conversions for cores within planetesimals of various sizes. For example, the cooling rates for IIIE irons could have been established at a 25-km depth in a 1,000-km-diameter body or a 100-km depth in a 200-km-diameter object, but even the center of a 100-km-diameter asteroid would have cooled faster. Cores buried to depths

Figure 7.1: This painting, by artist Mark Maxwell, illustrates the destruction of asteroids caused by impact.

greater than 180 km in a body 1,000 km in diameter would not cool to 500 °C (the approximate temperature at which diffusion of nickel in metal effectively ceases and the cooling rate is locked in) even within the age of the solar system, 4.56 billion years. The measured cooling rates for iron meteorites indicate that all their parent bodies must have been modest-

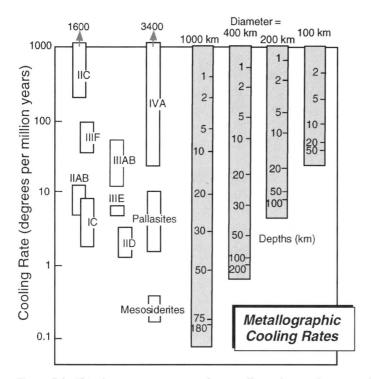

Figure 7.2: This diagram summarizes the metallographic cooling rates for various groups of iron and stony–iron meteorites. Because the scale is logarithmic, groups that fall in the upper part of the box (for example, IIC) show the greatest variation in cooling rate. The shaded vertical scales at the right illustrate depth-of-burial conversions for asteroids of different sizes that would produce these cooling rates. The cooling-rate data suggest that iron meteorites must have formed as cores in relatively small bodies or near the surfaces of somewhat larger bodies. The conversions do not take into account the insulation that would be provided by a surface regolith, so depth of burial and hence parent body sizes may have been even smaller. Cooling rates were determined by Henning Haack and Kaare Rasmussen (University of Odense, Denmark).

sized asteroids, especially if the irons formed within cores at the centers of these objects.

Henning Haack and his collaborators at the University of Odense have explored the thermal effects of regoliths on the surfaces of asteroids undergoing heating and differentiation. Powdered regolith is an effective insulator, holding in the heat. The presence of such a surface mulch retards the rate at which the core loses heat, so it may cool as much as ten times slower than a body without a regolith. Therefore a regolith-covered body could be smaller than a bald asteroid that has the same cooling rate.

These results strengthen the conclusion that most irons and pallasites formed within bodies that had diameters of less than 100 km.

The most straightfoward way to obtain samples of a central core is by catastrophic disruption of a differentiated asteroid. Alternatively, it might be possible to expose the core through erosion of the enclosing rocky mantle by a succession of somewhat smaller but still destructive impacts. Under these conditions a core might remain largely intact, perhaps like the rounded yolk of a hardboiled egg that breaks cleanly away from its white envelope. In such cases, the asteroidal relics would be bare metallic spheres, nearly devoid of adhering mantle silicates.

Each iron group represents a distinct chemical system, presumably the core of a different asteroid. Within any central core it seems reasonable that fractional crystallization would have produced chemical variations, which in turn would have produced mineralogical and structural differences, and indeed these are observed. However, one variation among members of the same iron group that would not have been expected is differences in cooling rate, because metal is such a good thermal conductor that the entire core should have cooled at nearly a uniform rate. As expected, most iron groups show only a modest range of cooling rates, and even these intervals may be due partly to analytical uncertainties. For example, samples of the largest and best-studied iron group, IIIAB, have metallographic cooling rates that vary within a narrow range of only approximately 70° per million years. In stark contrast are the highly diverse cooling speeds of group IVA irons, which cooled at rates varying from 30° to more than 3,000° per million years. Such large cooling-rate variations were once thought to imply that a few iron groups formed as small nuggets of metal distributed at various depths within an asteroid, analogous to raisins in a loaf of raisin bread. Each mass of metal would then have had its own unique cooling history, predicated on its burial depth. A particular iron meteorite group would contain samples from a number of these raisins, each with its own cooling rate, but the parent body would have stamped its unique chemical fingerprint on all these metallic segregations.

Iron groups with highly variable cooling rates, like IVA, are now usually attributed to impact disruption and reassembly. The differentiated IVA parent body, containing an already solidified but still hot core, was catastrophically broken into pieces that reaccreted within minutes or hours, before cooling of the metal fragments through 500 °C could occur. Huge amounts of silicate dust produced in the impact might have minimized heat loss by radiation from the larger iron fragments. In the resulting reaccreted rubble pile, still-hot chunks of the former core were mixed with mantle material at various temperatures and were buried to different depths, ensuring a wide range of cooling rates. A subsequent impact destroyed the rubble pile, allowing a handful of the iron fragments from different locations within the body to be sampled as meteorites.

The Boundary between Core and Mantle

Pallasites are thought to have formed at the outer fringes of cores where molten metal was in contact with the overlying mantle silicates. We have already seen that similarities in metal compositions suggest that the IIIAB iron core was rimmed by main-group pallasites. As shown in Figure 7.2, the metallographic cooling rates of pallasites were apparently slower (less than 10 °C per million years) than those of the associated IIIAB irons (faster than 10 °C per million years), a reversal of the expected relationship if pallasites coated the IIIAB iron core and thus were closer to the parent body surface. There are a number of possible explanations for this perplexing observation. Perhaps the cooling rate of the core slowed toward the end of its crystallization or the calculated cooling rates for pallasites are in error. Another possibility is that the chemical similarity between pallasites and IIIAB irons is fortuitous. The Eagle Station pallasites are not obviously related to a known iron group, so it is conceivable that their parent body may not have been eroded down to the point where raw metal core was exposed.

The close-packed arrays of olivine grains in pallasites indicate that they accumulated in this condition and liquid metal filtered in to occupy the spaces between the silicate crystals. If main-group pallasites rimmed the IIIAB core, metal infiltration happened after most of the metal core had already solidified because the pallasite metal has a more highly fractionated composition, plotting at the late-crystallizing end of the IIIAB iron trend on the nickel–iridium diagram.

Given the drastic differences in density between metal and silicates, it is not obvious how core and mantle materials became intermixed. The weight of a layer of olivine crystals accumulated from silicate magma at the core–mantle boundary might have exerted a downward force sufficient to submerge the lowermost olivines within molten metal, as illustrated in Figure 7.3. Alternatively, shock waves caused by impacts onto the surface of a differentiating body might have squeezed dense liquid metal upward into the overlying mantle, or cooling stresses might cause intermingling of olivines with molten metal. However it happened, the

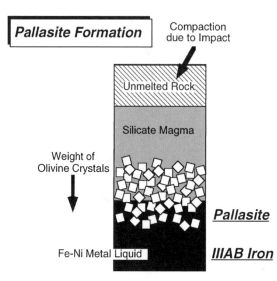

Figure 7.3: Because of differences in their densities, it is difficult to envision how molten metal and olivine crystals became mixed to form pallasites. These components would have been in direct contact at the boundary between the core and mantle. The lowermost olivine crystals that accumulated at the base of a magma might have been submerged into the core by the weight of overlying olivine grains. Alternatively, impacts into the surface or possibly stresses due to cooling might have squeezed metallic liquid into the mantle. The metal compositions for IIIAB irons and main-group pallasites are so similar that they are thought to have formed in the same parent asteroid.

occurrence of at least two distinct groups of pallasites indicates that this mixing process was not an isolated incident in only one core.

After mixing, the solidification of the metallic liquid must have been quick so as to prevent gravitational separation of silicate from metal. The tendency for metal and silicate to separate is demonstrated by the fact that most differentiated meteorites contain either less than 1% or greater than 99% metal by weight. Pallasites are unstable mixtures frozen in place. It seems incongruous that pallasites would have solidified rapidly but, once solidified, cooled slowly.

In order for core–mantle boundaries to have been sampled, pallasite parent bodies had to have been substantially eroded by impacts. Such bodies, if they survive today, would probably be similar to the metallic spheres discussed above, except that their surfaces would be studded with olivine crystals. If the main-group pallasites are related to IIIAB irons, as is commonly believed, their parent body no longer exists because the high abundance of IIIAB irons suggests that the Earth has received a representative sample of the underlying core. However, the parent asteroids of other pallasite groups could conceivably still be intact.

A Huge Mesosiderite Asteroid?

The intimate admixture of core and crustal materials, without incorporating fragments of the mantle rocks that would have comprised the intervening region in any plausible parent body, requires a complex history for the mesosiderite parent body. Presumably a naked iron core, stripped of its enclosing silicate by impacts, collided and mixed with the basaltic crust of another differentiated asteroid. This event may have involved disruption and reassembly into a rubble pile, as envisioned for some chondrite parent bodies. A clustering of argon-40 ages (which are commonly reset during impacts) for mesosiderites at approximately 3.9 billion years has been interpreted as an indication of a catastrophic collision at that time. However, the impact energy required for heating most of the asteroid by several hundred degrees and releasing argon gas might exceed the energy necessary to fragment the object.

Other workers hold the contrary view that the late argon-40 ages reflect the formation of mesosiderites in a very large asteroid, perhaps 400 to 800 km in diameter. In this case, deeply buried materials in a differentiated body would cool so slowly that argon gas would not begin to be retained until 0.6 billion years after the body formed. The extremely slow cooling rates determined from nickel diffusion in mesosiderite metal (approximately half a degree per million years) imply deep burial, which is possible only within a large asteroid.

A Cornucopia of Cores

It is difficult to say how many parent bodies are represented by our present collections of iron and stony–iron meteorites, but the number must be fairly large. Each of the thirteen groups of irons was possibly derived from a different core (although some workers favor lumping IAB and IIICD irons into a single group). Anomalous irons that are unique or have only a few recognized members expand the number of possible parent bodies to perhaps sixty. Also, two pallasite parent bodies have been sampled, although the main-group pallasites and the group IIIAB irons probably were derived from the same object. At least one mesosiderite planetesimal is also required.

This is a very generous number of asteroids, especially when compared with the limited number of stony meteorite parent bodies that have apparently been sampled. The explanation for the high abundance of sampled iron meteorite parent bodies may be deceptively simple, and apparently it lies in the fact that iron meteoroids are stronger than stony ones. As objects orbit in space, they are subjected continually to degradation by mutual impacts. Irons are more likely than stones to survive such encounters, and thus over time their parent bodies become better represented in meteorite collections. Another possible factor is the ease with which irons can be recognized among meteorite finds.

Many of the iron meteorite parent bodies that have already been sampled almost certainly no longer exist. Sampling cores is not like sampling the near-surface lava flows of asteroid 4 Vesta. That we have these core samples at all implies that their original parent bodies have been largely, if not totally, disrupted. Therefore, the chance of pinpointing the still-orbiting body from which a specific iron group was derived seems less likely than for stony meteorite parent bodies.

Shiny Beads

The most diagnostic information for the recognition of metallic asteroids is based on the reflection of signals beamed from radar facilities on the Earth (specifically the National Astronomy and Ionosphere Center's Arecibo Observatory in Puerto Rico and the Jet Propulsion Laboratory's Goldstone Radar in California). When the characteristics of the echo are compared with to those of the transmitted radar signal, it is possible to deduce the asteroid's properties. Because of the long wavelengths involved, radar is sensitive to the body's surface porosity as well as to its composition. Thus it is not possible to determine unambiguously the metal abundance, despite the fact that metallic objects are especially bright reflectors. Nevertheless, radar studies of several asteroids provide strong evidence of their metallic composition. Steven Ostro and his colleagues of the Jet Propulsion Laboratory have determined that the radar albedo of main-belt asteroid 16 Psyche is consistent with that of almost pure metal, with a surface regolith composed of fine, porous material. This asteroid, 260 km in diameter, is probably the largest exposed mass of refined metal in the solar system. Near-Earth asteroid 1986 DA is also metallic, a 2-km chunk of gleaming iron with hardly any regolith (see Figure 7.4).

The spectrum of sunlight reflected from the surfaces of iron meteorites is frankly rather boring. Although irons are highly reflective, their spectra exhibit no peaks and valleys, the diagnostic absorption features on which asteroid spectral interpretation has come to depend. The sloping, almost linear, spectra for four irons are illustrated in Figure 7.5. The only

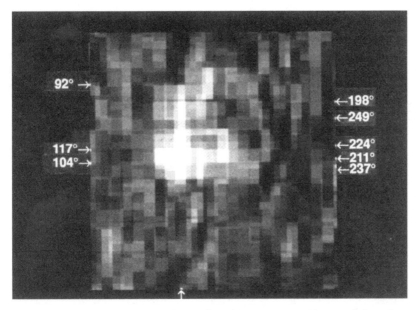

Figure 7.4: Eight radar images, obtained as the target asteroid rotated, have been superimposed to construct this composite image of near-Earth asteroid 1986 DA. The numbers at left and right refer to range locations of the echoes, and each square frame corresponds to approximately 5 km on a side. The very high radar reflectivity of this asteroid, as indicated by the bright spot, demands a surface dominated by metal. The radar image was obtained by Steven Ostro and collaborators (Jet Propulsion Laboratory).

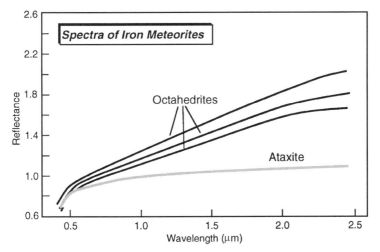

Figure 7.5: The reflectance spectra of three octahedrites have reddened slopes but are otherwise rather featureless. The ataxite spectrum is flatter, which is characteristic of iron meteorites with higher nickel contents. These spectra are of limited use in recognizing iron meteorite parent bodies by remote sensing.

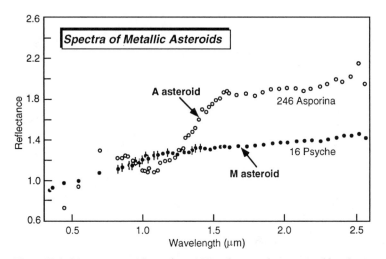

Figure 7.6: M-type asteroids, such as 16 Psyche, are characterized by sloping spectra that resemble those of iron meteorites. In contrast, A-type asteroids exhibit a prominent absorption band at 1 μm, characteristic of olivine. Asteroids like 246 Asporina are likely candidates for pallasite parent bodies. These spectra were obtained by Jeffrey Bell and co-workers (University of Hawaii).

real variation within these spectra is a decrease in slope with increasing nickel content. This is demonstrated by the relatively flat spectrum of the ataxite, containing nearly 12% nickel, relative to those of the other three octahedrites.

However unexciting these spectra may be, their high albedos, coupled with the absence of absorption features attributable to silicates or opaque materials like organic matter, may be at least partly diagnostic for metal-rich planetesimals. One class of asteroids, called the M type, exhibits this kind of sloping, featureless reflectance spectrum (see Figure 7.6), and these objects are commonly thought to be iron cores that have been stripped of their mantles. Both 16 Psyche and 1986 DA, previously identified as metallic by their radar reflectivity, are classified as M asteroids. However, the M group must be diverse, as some of its members have radar returns that can be matched by mixtures of silicate and iron metal, possibly like enstatite chondrites or group IAB irons with abundant silicate inclusions. A few M asteroids show a 3-μm absorption feature, suggesting the presence of clays, which is inconsistent with a metallic composition.

Incompletely denuded planetesimals or rubble pile asteroids consisting partly of chunks of core material would contain silicates in addition to metal. Such bodies would possibly have spectral signatures much like

those of S-type asteroids. In Chapter 3 it was suggested that some S asteroids may be ordinary chondrite parent bodies, but a combination of
metal and silicate from differentiated asteroids may provide an equally
acceptable model. It is difficult to specify the proportion of metal in S
asteroids because of the featureless spectrum of this material.

The reflectance spectra for stony–iron meteorites are very difficult to
obtain in the laboratory, because it is virtually impossible to crush mixtures of metal and silicate uniformly. It may seem curious to worry about
crushing malleable metal for this measurement; however, metal would be
brittle at the frigid temperatures of interplanetary space, and its behavior
when it experiences an impact on its parent body would probably differ
significantly from that in a warm laboratory on Earth. Because of this, any
comparison of measured spectra for asteroids with stony–iron meteorites
should be viewed with caution. A better approach is probably to calculate
what the spectra would look like by numerically combining the spectra
of a powdered octahedrite and either olivine (for pallasites) or a eucrite
(for mesosiderites).

Some very respectable matches for these calculated stony–iron reflectance patterns have now been discovered among asteroids. Potential parent bodies for pallasites have been identified as A-type asteroids. An example of an A-asteroid spectrum, 246 Asporina, is compared with the
spectrum of the M asteroid 16 Psyche in Figure 7.6. A asteroids exhibit
strongly sloping spectra with a strong absorption band at 1 μm (characteristic of olivine) and no absorption feature at 2 μm (characteristic of
pyroxene). Althought the spectrum of olivine is unambiguous, it is difficult to specify exactly how much metal is present in these bodies. Some
of them may be olivine achondrites without much metal. Currently recognized A asteroids have diameters of 30 to 65 km.

The mixing of metal with silicates seems almost accidental in mesosiderites and likely occurred when an iron meteoroid collided with a Vesta-
like surface. It has even been suggested that 4 Vesta itself might be the
object from which mesosiderites were derived. Although the silicate fractions of mesosiderites are very similar to those of HED achondrites, there
are several reasons to believe that this asteroid is not the mesosiderite
parent body. The rotational spectral map of Vesta does not disclose any
highly reflective regions rich in metal. Also, the very slow cooling rates
of mesosiderites argue for deep burial, so Vesta would have to have been
destroyed to liberate the mesosiderites.

The identification of possible asteroidal parent bodies for iron and
stony–iron meteorites permits some inferences to be drawn about their

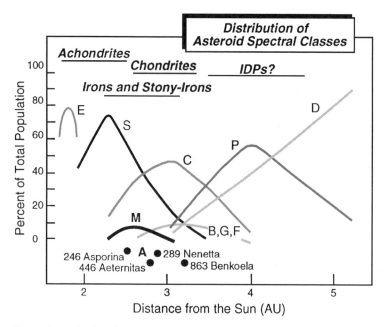

Figure 7.7: The distributions of various spectra classes of asteroids indicate that plausible parent bodies of irons (M asteroids) and stony–irons (A and possibly some S asteroids) are located in the inner belt, mostly between 2 and 3 AU. Did all the bodies in this inner region partially melt and differentiate? The A asteroids are too rare to be shown as a percentage of the total asteroid population, but the approximate orbital distances of four of them are illustrated.

occurrence in space. Figure 7.7 shows the distributions of M asteroids relative to other spectral types in the main belt. Also shown as dots are the positions of four A-type asteroids; these bodies are uncommon and thus would not define a significant hump in this diagram. If irons are derived from M asteroids and stony–irons from A or possibly some S asteroids, their clustering in the inner belt is consistent with the previous inference from the distribution of achondrite parent bodies that differentiated objects are concentrated in the inner main belt.

Asteroid Families

To view the internal structure of an apple, we must first cut it open. Nature has already broken open asteroids for our perusal, but the job may have been done too well. The destruction that occurs when planetesimals collide catastrophically produces fragments with a great variety of sizes. It is likely that one of the impacting bodies would have been much smaller than the other, and cratering studies show that the smaller projectile experiences far greater stresses than the larger target body. The resulting pulverization or melting of the smaller projectile means that only one asteroid (the larger body) forms the majority of observable fragments resulting from the collision.

Reconstructing the original target planetesimal from these dispersed fragments is like putting together a puzzle face down, with no visual clues from the emerging picture on the front side. However, in a series of papers beginning in 1918, Japanese astronomer Kiyotsugu Hirayama found a way to do just that. Hirayama surmised that the breakup of an asteroid into a collection of fragments, which he called a family, would result in similar orbital characteristics for these bodies. In practice, corrections must also be made for minor orbital perturbations caused by nearby Jupiter. This results in what are called proper orbital characteristics. On the basis of similar proper orbital features, Hirayama was able to recognize clusters of asteroids, which we now refer to as **Hirayama families**. He hypothesized that the members of any one family were collisional fragments of the same original planetesimal.

Hirayama's families, consisting of the most obvious orbital groupings of asteroids, have stood the test of time, but other less conspicuous families added in later years are more problematical. One observation that supports the reality of asteroid families produced during collisions is that orbital velocity differences among family members are small, generally no more than a kilometer per second. These speeds are only a tiny fraction of their total orbital velocities and must approximate the original dispersal velocities when their parent bodies fragmented. Artificial satellites that have exploded in space produce fragments with velocity distributions similar to those of Hirayama families. The effect of these small velocity differences over time is to spread out the family members along the original planetesimal's orbit, something like the dust trail left by a comet.

It has long been recognized that asteroid families potentially provide a way to examine the diverse materials that comprise differentiated bodies. Surveys of the reflection spectra of the members of a particular family should reveal the compositional variability within the original body. Unfortunately, most of the accepted asteroid families have members that are uniformly chondritic in composition. For example, all sixty-one asteroids comprising the the Koronis family, named for 158 Koronis, are

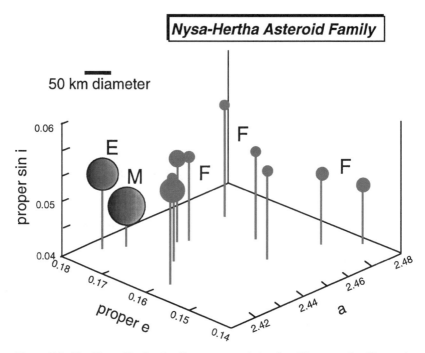

Figure 7.8: The Nysa–Hertha family was one of the first Hirayama families to be recognized. This diagram, adapted from the work of Clark Chapman and co-workers (Planetary Science Institute), shows the clustering of proper orbital elements for this group (a is the semimajor axis of the orbit, e is the eccentricity or degree of oblateness of the ellipse, and $\sin i$ is the sine of the angle of inclination of the orbit to the ecliptic). The relative sizes of family members are indicated by the sizes of the circles. 44 Nysa is the E asteroid and 135 Hertha is the M. It was originally thought that this family represented a fragmented enstatite achondrite asteroid, with Hertha as the core and Nysa as a large surviving piece of the mantle. However, subsequent spectral characterization of the small family members has revealed that they are all of the rare F type, thought to be similar to carbonaceous chondrites. Although the smaller bodies may be a disrupted asteroid, Nysa and Hertha are probably interlopers.

S-type objects that were originally part of a larger, homogeneous body. The families that contain metallic asteroids tend not to make much sense as reconstructed objects. A notable example is the Nysa–Hertha family (see Figure 7.8), named for 44 Nysa (an E asteroid) and 135 Hertha (an M asteroid). These two asteroids, the largest bodies within a family of seventy-seven objects, could plausibly have once fitted together as core and silicate mantle, with the remaining small chunks being more highly fragmented relics of the mantle and crust. This interesting hypothesis was disproved, however, when a spectral survey revealed that all the smaller family members analyzed were of the F type, thought to be related to carbonaceous chondrites. The rarity of F asteroids elsewhere in the belt implies that the smaller bodies in this grouping do constitute a true family, but Nysa and Hertha are probably interlopers. Reconstructions of other families containing M asteroids are similarly disappointing.

Although asteroid families apparently have not offered glimpses of the otherwise unobservable interiors of differentiated bodies, as originally hoped, spectral and radar imagery of smaller objects may yet provide this kind of insight. Older families gradually disperse, perhaps accounting for the absence of a differentiated family grouping. Given the occurrence of a number of chondritic families, however, it is surprising that at least one core with its attendant mantle fragments has not been recognized.

Heavenly Irons

The hieroglyphic symbols shown in Figure 7.9 have been observed in a number of Egyptian pyramids. A literal translation of these ancient markings is "heavenly iron." Pharaohs were sometimes buried with artifacts constructed from iron meteorites, and it is doubtful that the desire to retain ownership of such objects in the afterlife stemmed solely from their utility or decorative appearance. From the name given to these metallic chunks it is clear that the ancient Egyptians recognized the uniqueness of their source and attributed some importance to it.

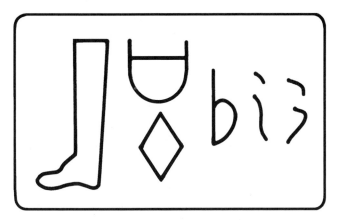

Figure 7.9: These hieroglyphic symbols, found in Egyptian pyramids, mean heavenly iron. For many years they appeared on the cover of the journal of the Meteoritical Society, an international organization devoted to research on meteorites, other extraterrestrial materials, and impact craters.

In this chapter we have attempted to fix the extraterrestrial sources of iron and stony–iron meteorites more precisely than did the Egyptians, but with only limited success. Iron meteorites clearly formed the cores of differentiated asteroids, and sampling of such deep-seated regions probably required the impact destruction of their parent bodies in most cases. Stony–irons, whether pallasites or mesosiderites, cooled slowly in the deep interiors of asteroids that must have been similarly disrupted. Several spectral classes of asteroids are plausible candidates for fragmented or partly denuded cores, but the featureless spectrum of iron metal makes quantitative estimation of metal contents impossible. Radar imagery offers a more definitive test for metallic objects, but the target objects must be large or in near-Earth orbits for their echoes to be recorded. The apparent restriction of metallic asteroids to the inner main belt is consistent with the inference from the distribution of achondrite parent bodies that melting and differentiation decreased away from the Sun.

The asteroidal sources of iron and stony–iron meteorites remain something of a mystery, as indicated by the relative brevity of this chapter. Regardless, their importance as indicators of otherwise hidden processes occurring in the very hearts of planetary bodies remains unchallenged.

Suggested Readings

The topic of iron meteorite parent bodies has been addressed infrequently in the scientific literature and almost never in popular papers or books. Most of the references below make challenging reading but are the only sources available.

GENERAL

Wasson J. T. (1985). *Meteorites: Their Record of Early Solar System History*, Freeman, New York.

Chapter IV of this book describes the evidence for and against core formation in asteroids.

RADAR AND SPECTRAL REFLECTANCE IMAGERY

Cloutis E. A., Gaffey M. J., Smith D. G. W., and Lambert R. J. (1990). Reflectance spectra of "featureless" meteorites and the surface mineralogies of M- and E-class asteroids. J. Geophys. Res. **95**, 281–293.

A detailed study of the spectra of iron meteorites provides support for the idea that irons are derived from M asteroids.

Gaffey M. J., Burbine T. H., and Binzel R. P. (1993). Asteroid spectroscopy: progress and perspectives. Meteoritics **28**, 161–187.

An authoritative and especially readable review article presenting representative spectra for M and A asteroids and discussing their compositional interpretation.

Ostro S. J., Campbell D. B., Chandler J. F., Hine A. A., Hudson R. S., Rosema K. D., and Shapiro I. I. (1991). Asteroid 1986 DA: radar evidence for a metallic composition. Science **252**, 1399–1404.

The most convincing evidence for the existence of a metallic asteroid, in this case one in near-Earth orbit.

PARENT BODIES FOR IRONS

Scott E. R. D. (1979). Origin of iron meteorites. In *Asteroids*, T. Gehrels, ed., University of Arizona, Tucson, AZ, pp. 892–925.

A thorough treatment of core formation in asteroids.

Haack H., Rasmussen K. L., and Warren P. H. (1990). Effects of regolith/megaregolith insulation on the cooling histories of differentiated asteroids. J. Geophys. Res. **95**, 5111–5124.

These calculations explore the thermal effects of insulating regolith and provide an argument for relatively small parent bodies for iron meteorites.

Taylor G. J. (1992). Core formation in asteroids. J. Geophys. Res. **97**, 14,717–14,726.

This technical paper describes the evidence that metallic cores formed in asteroids.

PARENT BODIES FOR STONY–IRONS

Scott E. R. D. (1977). Formation of olivine-metal textures in pallasite meteorites. Geochim. Cosmochim. Acta **41**, 693–710.

This paper describes various processes that could lead to mixing of liquid metal and silicate at core–mantle boundaries.

Haack H., Scott E. R. D., and Rasmussen K. I. (1996). Thermal and shock history of the mesosiderites and their large parent asteroid. Geochim. Cosmochim. Acta **60**, 2609–2619.

A technical summary describing the slow metallographic cooling rates and limited shock metamorphism of mesosiderites, leading to the inference that they formed in a large body.

ASTEROID FAMILIES

Bell J. F. (1989). Mineralogical clues to the origins of asteroid dynamical families. Icarus **78**, 426–440.

A technical paper containing tables listing the sizes and properties of the various proposed asteroid families.

Chapman C. R., Paolicchi P., Zappala V., Binzel R. P., and Bell J. F. (1989). Asteroid families: physical properties and evolution. In *Asteroids II*, R. P. Binzel, T. Gehrels, and M. S. Matthews, eds., University of Arizona, Tucson, AZ, pp. 386–415.

This chapter summarizes the compositions and masses of proposed asteroid families and describes the controversies surrounding these groupings.

8 A Space Odyssey

There is an anecdote, well worn with many variations, in which a grizzled denizen of rural Maine or some such place is asked to provide directions to a local landmark. After a number of aborted attempts, he is finally forced to the conclusion that "you can't get there from here." The very fact that we have meteorites on Earth shows that this punch line cannot apply to their parent bodies. Nevertheless, the difficulties and the complexities involved in the extraction of samples from asteroids or planets and their delivery to Earth would wither the resolve of most people trying to give explicit directions.

In previous chapters we have considered the characteristics of various kinds of meteorites and attempted to explore the nature of their parent bodies. The intent of this chapter is to tie up one monumental loose end – how these meteorites got from their parent bodies to the Earth. The delivery of meteorites is, of course, only a minor by-product of the clockwork of the solar system. These celestial workings are vastly more intricate than the working parts of timepieces, and, for small objects like meteorites, they cannot be observed directly. However, the mechanisms by which samples are liberated from their parent objects and the routes by which they find their way to Earth can be analyzed indirectly. It is not possible to specify precisely the entire odyssey of any particular meteorite, although the orbits at the times of impact are known for a handful of chondrites. As we see below, though, plausible paths from parent bodies to Earth can be reconstructed.

Asteroidal Traffic Accidents

Typical relative velocities for asteroids encountering other asteroids are of the order of 5 km/s. We cannot recreate in laboratory experiments exactly what happens when planetesimals collide at these speeds. We can determine the disastrous outcomes only by extrapolating the results of fragmentation experiments at much lower velocities, but studies of nuclear weapons tests provide some basis for believing that these calculations are correct.

It probably comes as no surprise that such accidental collisions liberate meteoroids from their parent bodies. Most of the energy of the original projectiles gets transformed into heat or does the work of asteroid disruption; however, as much as 10% of the original energy in cratering events is transmuted into energy of motion for the resulting debris. This kinetic energy provides a way for fragments to be cast free of their asteroidal moorings. If the energy of motion for a fragment exceeds the gravitational pull that binds the planetesimal together, the chip is said to have achieved **escape velocity**, and an independently orbiting meteoroid is born.

The brecciated nature of many meteorites demonstrates the role that impact processes have played in asteroidal evolution. We have even surmised that many meteorite parent bodies are reaccreted piles of rubble produced during large impacts. The densities of some asteroids, such as 253 Mathilde (see Figure 8.1), are so low (estimated at 1.3 grams per cubic centimeter in this case, not much different from water) that they must contain appreciable pore space, as appropriate for rubble piles but not for solid rock. Images of the cratered surfaces of asteroids and of the tiny moons of Mars, which are probably captured asteroids, also show the scars of numerous impact events in their past. Every planetesimal in the asteroid belt has suffered impacts at some scale over the 4.6 billion years of its

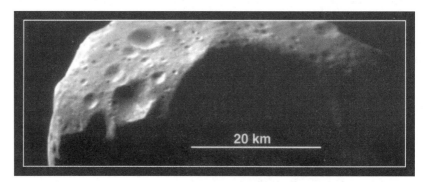

Figure 8.1: Asteroid 253 Mathilde, like other asteroids that have been examined at close range, has survived a barrage of impacts. This image from the *NEAR* spacecraft, taken at its closest approach to Mathilde in June 1997, shows abundant craters as small as 500 m across. The impact that produced the large (20 km in diameter and 10 km deep) crater must have nearly disrupted this asteroid, which has a mean diameter of 52 km. Impacts provide the mechanism for liberating meteorites from their parent bodies. Photograph courtesy of the Applied Physics Laboratory of Johns Hopkins University and NASA.

existence, and much of the present belt may now simply be battered colli-sional remnants. As a consequence, there have been ample opportunities for meteoroids to have been extracted from asteroids by mutual collisions.

Impact Ages of Meteorites

The timing of impact events on asteroids can be determined from gas-retention ages (introduced briefly in Chapter 2). ^{40}Ar (an isotope of argon produced by radioactive decay of ^{40}K, an isotope of potassium) is a gas readily lost during impact. Thus a cratering event will reset this isotopic clock, and the elapsed time measured by the accumulation of ^{40}Ar repre-sents the interval since impact flushed gas from the rock. Gas-retention ages can also be determined from ^{4}He, an isotope of helium that forms by the decay of uranium. Sometimes other radiogenic isotopic systems that do not involve gaseous isotopes are also reset, but usually this occurs in collisions only between large bodies.

The ^{40}Ar ages of HED achondrites indicate resetting within a relatively narrow time interval of 3.4 to 4.1 billion years ago. These ages are similar to the impact ages of lunar rocks, suggesting that asteroid 4 Vesta experi-enced the same kind of bombardment that the Moon did and implying that this cataclysm may have affected the whole inner solar system. It is doubtful that the HEDs in our meteorite collections were ejected from their parent body at this time, however, because small pieces cannot sur-vive in space for billions of years.

The gas-retention ages for ordinary chondrites identify collisional events that occurred less than 600 million years ago. For example, data for L-group chondrites form a pronounced cluster. This was originally interpreted to reflect an age of approximately 500 million years, but gas-retention ages based on ^{4}He have recently been readjusted to give an age of approximately 340 million years. This measurement, plus the obser-vation that L chondrites tend to be heavily shock metamorphosed, have been interpreted as evidence that the L-chondrite parent body was de-stroyed by impact at that time. The H and LL chondrites show ranges of ^{40}Ar and ^{4}He ages, implying a series of smaller, less disruptive impacts extending to recent times. These events may have liberated chondritic samples from their asteroidal sources.

The Properties of Orbits

Ejection of pieces of rock or metal from their asteroidal parent bodies is the crucial first step in the long trek to Earth. However, completion of this trip requires that the dislodged meteoroids be placed into orbits that intersect the orbital path of their eventual planetary target. In order to understand how this hurdle is surmounted, we must first examine the properties of meteoroid orbits.

The trajectory followed by the original asteroidal parent body traces out an ellipse, although it may look superficially like a circle. A similar elliptical path is followed by any smaller object revolving about a more massive body, whether it is a large planet orbiting about the Sun or a tiny spacecraft orbiting about the Earth. The three most important orbital elements that define the size, shape, and orientation of the ellipse are its **semimajor axis** (the greatest distance from the center of the ellipse to its periphery), its **eccentricity** (a measure of its departure from circularity), and its **inclination** (the angle between the orbital plane and the orbital plane of the Earth). The orbital **period**, the time required for completion of one orbital loop, is another useful parameter.

Although the orbital behavior of a small body may be described approximately by the geometry of an ellipse, the situation is actually somewhat more complicated. Often the gravitational effects of a third body also come into play. For example, the orbit of a spacecraft revolving around the Earth may be modified at some points by gravitational tugs exerted by the Moon. These **perturbations** result in minor sinuosities being superimposed on an otherwise perfect elliptical track. Unfortunately, whereas we have equations that specify the geometry of an elliptical orbit for the two-body problem, there is no general mathematical solution that describes the motions when three bodies are involved. However, these perturbations can be predicted by computer simulation. The calculation procedure is tedious, involving many small steps. The forces are determined at each step, and the small body of interest is repositioned (perturbed) according to the results of the previous step.

For a fragment of an asteroid to become a meteorite, its orbit must be perturbed in such a way that it crosses the orbital path of the Earth. For objects derived from the asteroid belt, this requires a significant increase in eccentricity. This is all well and good, except that the impact events that freed meteoroid fragments from their asteroidal parent bodies in most cases cannot have altered their orbits this drastically. As a consequence of collisions, asteroidal fragments are usually ejected at velocities of perhaps a few hundred meters per second, too small for causing the drastic eccentricity changes required for achieving planet-crossing orbits. Recall that in a previous discussion of asteroid families, we learned that most of the debris from a collision spreads out along the orbital path of the original parent object. Collisional shocks capable of producing more than small, say 0.1 AU, changes in a semimajor axis would probably also grind the liberated fragments to dust. The highly elliptical meteoroid orbits required for Earth capture must have been produced by some other kind of perturbing force. But what kind of force could have done this?

Geography of the Asteroid Belt

The positions of planetesimals within the main asteroid belt offer an important clue to the identity of this perturbing force. The distribution pattern of asteroids is anything but smooth. This is clearly shown in Figure 8.2, a plot of the semimajor axes of the orbits of the first 400 numbered asteroids. This snaggletoothed distribution pattern was first noted by the astronomer Daniel Kirkwood in 1867, and the relatively empty slots within the asteroid belt are now called **Kirkwood gaps**. These are not actually voids, because asteroids with elliptical orbits pass through them. However, the gaps are regions in which almost no asteroids remain for long.

Although the Kirkwood gaps seem to occur at random distances from the Sun in Figure 8.2, they are in fact at very specific locations relative to the orbit of Jupiter. Each of these localities corresponds to a **mean-motion resonance**, defined as an orbit whose period corresponds to a simple fraction of Jupiter's orbital period. For example, an asteroid whose orbital period is one-half that of Jupiter will find itself adjacent to the giant planet on every second revolution around the Sun. In Figure 8.2, mean-motion resonances are marked by appropriate fractions. The numerator in each fraction refers to the number of revolutions an asteroid at that position must make in order to line up with Jupiter after it has made the number of revolutions given by the denominator. The correspondence between the positions of mean-motion resonances and Kirkwood gaps indicates a causative link.

Because asteroids in resonant positions experience close encounters with Jupiter more often than do asteroids in other, random locations, they also feel the powerful pull of Jupiter's gravity more frequently. These tugs cause objects in these regions to undergo rapid and irregular variations in their orbital elements. Over time, these perturbations can eliminate asteroids from the Kirkwood gaps. In particular, Jack Wisdom at the Massachusetts Institute of Technology has shown that asteroids with semimajor axes of approximately 2.5 AU, where the Kirkwood gap corresponding to the 3/1 mean-motion resonance with Jupiter is located,

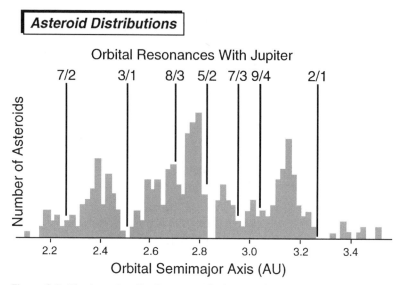

Figure 8.2: The irregular distribution with distance from the Sun (expressed as the orbital semimajor axis) of the first 400 numbered asteroids is clear from this diagram. Vertical lines indicate vacancies, called Kirkwood gaps, in the main asteroid belt that are due to mean-motion resonances with the orbit of Jupiter. The fraction 7/2 means that an asteroid in that position makes 7 revolutions around the Sun for each 2 revolutions completed by Jupiter. Asteroids at such resonant positions are affected by Jupiter's massive gravitational field more often than asteroids in other, random positions. In this way, objects temporarily stored in these gaps can be perturbed into Earth-crossing orbits.

develop very chaotic motions, resulting in strong eccentricity increases on time scales of only a million years.

Besides mean-motion resonances, another important source of perturbations exists. **Secular resonances** likewise can alter the orbits of meteoroids so that they become Earth crossing. In the case of secular resonances (usually identified by the symbol v), what matters are not the orbital periods, but rather the times over which the orbits change their mutual orientation (precess) because of planetary perturbations. For example, the v_6 secular resonance affects bodies whose orbits precess at the same rate as Saturn. The locations of secular resonances vary with inclination as well as with the semimajor axis, as illustrated by the curves in Figure 8.3. The v_6 resonance forms a curved upper boundary in inclination for most asteroids in the main belt. Other important secular resonances, such as v_{16}, define additional depopulated regions. On the left-hand side of this figure, the 4/1 mean-motion resonance and the v_6 and the v_{16} secular resonances combine to define the inner edge of the main asteroid belt.

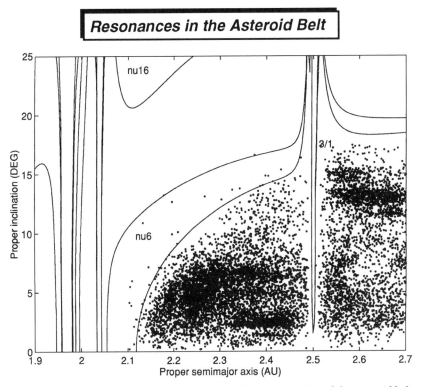

Figure 8.3: This detailed map of resonances in the inner portion of the asteroid belt illustrates a second kind of escape hatch. Each tiny dot represents the location of an asteroid. The horizontal axis indicates distance from the Sun (as in Figure 8.2), and the 3/1 mean-motion resonance at 2.5 AU is shown. The vertical axis indicates the inclination of the asteroids' orbits relative to the ecliptic plane. Proper semimajor axes and inclinations are the average values of these orbital elements after periodic planetary perturbations have been filtered out. Most of the curves in this diagram are secular resonances, gaps in asteroid distribution produced when the rates of orbital precession overlap those of the giant planets. The ν_6 and ν_{16} (pronounced nu and labeled here as nu6 and nu16) secular resonances are identified. Note that the inner edge of the asteroid belt occurs where many of these resonances cluster. Asteroids located close to or inside the resonant strips are the most likely candidates for meteorite parent bodies. Figure courtesy of Paolo Farinella (University of Pisa, Italy).

Escape Hatches

Resonances provide very effective escape hatches for most of the meteoroids exiting the asteroid belt. Fragments that are liberated at relatively low velocities by asteroid impacts may wander into nearby resonances, where their orbits are perturbed so that they enter the inner planet region. The 3/1 mean-motion resonance and the ν_6 secular resonance, in

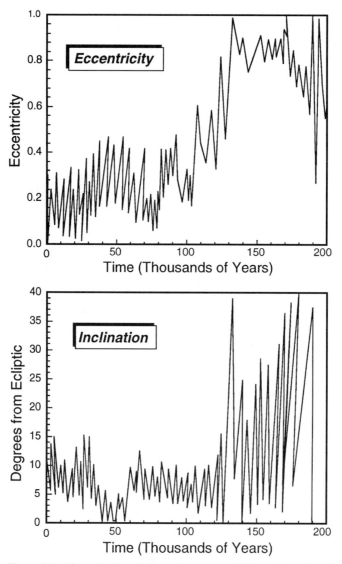

Figure 8.4: The orbital evolution of a meteoroid injected into the 3/1 mean-motion resonance can be studied by numerical simulation. Each object behaves differently, but the one shown here illustrates a typical outcome. At the beginning of the simulation (left-hand side of the figures), the fragment is at 2.5 AU (the location of the 3/1 resonance) with eccentricity and inclination equal to zero. Chaotic interactions with Jupiter's gravity field cause these orbital elements to oscillate, and after a million years the eccentricity becomes so high that the object is Earth crossing. It remains in a highly eccentric orbit while its inclination oscillates wildly, until eventually it hits the Earth (or Mars) or is driven out of the solar system by a close planetary encounter. This representative computer simulation is one of many performed by Paolo Farinella and colleagues (University of Pisa, Italy).

particular, are thought to make significant contributions to the meteorite flux reaching the Earth.

Paolo Farinella and his colleagues of the University of Pisa, Italy, have calculated the probabilities that fragments from specific asteroids might be injected into a resonance in their neighborhood. They concluded that a handful of fortuitously situated asteroids are probably the parent bodies for most meteorites. For example, asteroid 6 Hebe is located so close to the ν_6 secular resonance that a large fraction of its ejecta should end up within this chaotic region. Their simulation indicated that four of eighteen fictitious fragments ejected from Hebe, which were described in Chapter 3 as a plausible parent body for the H-group ordinary chondrites, became Earth crossing. One example, illustrated in Figure 8.4, shows that the eccentricity of one fragment is pumped up so that it can intersect the Earth's orbit within a million years, after which time its inclination oscillates wildly until it impacts the planet. These calculations demonstrate that fragments from Hebe can travel to Earth and support the postulated connection between this asteroid and a specific meteorite class.

Unlike Hebe, asteroid 4 Vesta is located far from resonances, making it more difficult for meteoroid fragments to be ejected into chaotic regions where they can dynamically evolve into Earth-crossing orbits. Richard

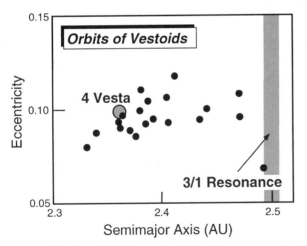

Figure 8.5: Asteroid 4 Vesta is located far from a resonant escape hatch, but a bridge of small asteroids with spectra similar to those of Vesta link this asteroid to the 3/1 mean-motion resonance. This diagram shows the proper semimajor axis and inclination (expressed as the sine) of the orbits of these objects, with the 3/1 resonance occurring at 2.5 AU. The vestoids form a family of objects clustered mostly around Vesta itself, but with a tail extending into the escape hatch. This diagram is based on observations by Richard Binzel and Shui Xu (Massachusetts Institute of Technology).

Binzel and Shui Xu of the Massachusetts Institute of Technology have studied the spectral properties of twenty faint, small asteroids (generally 5 to 7 km in diameter) in Vesta's vicinity, all of which resemble the spectra of HED meteorites. These objects must be fragments ejected from Vesta during a large impact. The 250-km-diameter crater seen in *Hubble Space Telescope* images of Vesta is certainly large enough to have produced these small asteroids. The small Vesta-like objects (vestoids) occupy orbital positions that bridge the gap between Vesta and the 3/1 mean-motion resonance (see Figure 8.5). To reach these locations, the objects must have been ejected with velocities greater than 500 m/s, much faster than had previously been thought possible. The discovery of this bridge of HED-like planetesimals reinforces the conclusion that Vesta is the HED parent body, as well as the hypothesis that resonances are escape hatches for materials derived from main-belt asteroids.

The Planetary Prison

The kinetic energy requirements for liberating fragments from asteroids are really quite modest. In fact, some future astronaut standing on the surface of a small asteroid could possibly launch a rock to escape velocity by using no more than a baseball throw. Large bodies, however, are a different story. Although large moons and planets experience the same buffeting by impacts that asteroids do, their massive gravity fields exert a near stranglehold on the fragments produced during cratering. It was once supposed that any smaller chips that might be ejected from planets by impacts would have experienced such high degrees of shock that they would be pulverized, melted, or even vaporized. Yet no other natural means of meteorite ejection seems possible. The energy of rapidly expanding gases during volcanic eruptions is far too small to accelerate fragments to planetary escape velocities, and other geologic phenomena are even less capable launching mechanisms. How, then, have we received intact pieces of the Moon and probably Mars?

The answer, at least in part, is provided by calculations in which differences in the impact behaviors of target materials near the ground surface versus those at depth are considered. The disturbance created by an impacting object propagates through the subsurface as a stress wave whose force lessens as it moves away from the impact site, like an expanding wave produced when a pebble is thrown into a pond. Target rocks close to the impact site are melted and broken into dust or small grains, with the fragment size increasing away from ground zero, as illustrated in Figure 8.6. However, rocks very near the ground surface experience

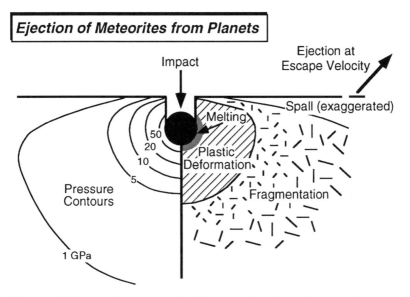

Figure 8.6: This sketch schematically illustrates the effects of a major impact on a planetary surface. The impacting object, of course, will be pulverized or vaporized. The left half of the figure shows contours of shock pressures (in gigapascals) experienced at various distances from the impact site. The right-hand side illustrates the effects produced in the target rocks. Close to the impact site, melting and pulverization take place, but farther away the rocks are broken into larger fragments. At the ground surface, virtually unshocked fragments spall off and can be accelerated to planetary escape velocities. This diagram is based on calculations by Jay Melosh (University of Arizona).

several kinds of shock waves that partially cancel each other. This area of wave interference offers some shelter from the full force of the shock wave. Calculations indicate that some of this near-surface material will spall off as relatively unshocked fragments that can be accelerated to high speeds. Limited shock is important in the case of lunar meteorites and nakhlites, which have experienced only minor shock metamorphism.

The only potential problem with this explanation is that the chips ejected at planetary escape velocities should be rather small. The lunar meteorties are generally the size of golf balls, and calculations indicate that rocks of this size could readily escape from the Moon's gravitational grasp (requiring a speed of 2.4 km/s). Some Martian meteorites are larger, however, up to grapefruit sizes, and these are thought to have lost perhaps half their original mass during transit through the Earth's atmosphere. The largest craters in the young terrain of Mars, a necessary location from which to derive most of these young meteorites, are approximately 30 km in diameter. The sizes of the fragments that could be accelerated to

Martian escape velocity (approximately 5 km/s) from such craters are approximately 1 m, so large craters are apparently capable of launching small rocks from planetary surfaces.

The precise mechanisms by which rocks escape from planetary gravity fields are not well understood. Computer simulations (sometimes called Monte Carlo simulations, an allusion to the statistical nature of these calculations) allow the orbital evolution of individual particles in space to be tracked. Once free of the gravitational hold of the Moon, a lunar meteorite may go into orbit about the Earth, eventually spiraling into our planet within a few decades. Most lunar ejecta, however, will be propelled into elliptical orbits about the Sun. These objects periodically reapproach the Earth and on close approaches may be scattered into new orbits by the Earth's gravitational field. Some of these Sun-orbiting bodies will fall to Earth, usually within a million years or so, with the remainder eventually being driven into the Sun or ejected from the solar system.

Rocks launched from Mars take significantly longer to reach the Earth than do lunar rocks, simply because their orbits do not initially cross that of our planet. These objects, like those in the asteroid belt, are subject to resonances related to other planets and, over a few millions of years, their orbits are perturbed so as to become Earth crossing. Most that fall to Earth will do so within 10 million years or so, with much of the remainder eventually being driven into the Sun.

Meteorites Exposed

When stimulated by light, a chemical reaction on plastic film produces a photographic negative. In a somewhat analogous manner, meteorites orbiting in space act as film when exposed, in this case to cosmic rays. The reactions resulting from this irradiation provide an indelible record of their travels.

When cosmic rays strike the nuclei of atoms that comprise orbiting pieces of rock or metal, they tend to dislodge protons and neutrons, a process known as **spallation**. These reactions change the identities of the target atoms; for example, the removal of one proton and two neutrons from ^{56}Fe (an isotope of iron) produces ^{53}Mn (an isotope of manganese). In some cases the newly created isotopes are sufficiently different from what already comprises the object that they can be recognized and analyzed. Spallation-produced isotopes can be either stable or radioactive, but the radioactive ones are much easier to measure.

Cosmic rays are particles with a variety of energies. Those from solar wind and flares are fairly weak and cannot bore deeply into rock or metal, but more energetic galactic cosmic rays can penetrate to depths of approximately a meter. Thus only the outermost meter of large meteoroids, really negligible amounts of material, will experience spallation reactions. Put another way, most orbiting meteoroids will be protected from cosmic irradiation until they are broken into fragments of meter size or smaller. Whenever this happens (by impacts), they begin to accumulate spallation products. The time interval calculated from measurements of the accumulated radionuclides produced by cosmic-ray exposure, called the exposure age, then represents how long a meteorite traveled in space as a meter-sized or smaller object. Exposure ages are in most cases only approximations, probably accurate within a factor of two, because so many assumptions are involved in calculating them. Nevertheless, they provide otherwise unobtainable information on the lifetimes of small meteoroids orbiting in space.

A complication is that cosmic irradiation can also occur on the surfaces of asteroidal parent bodies before smaller meteoroids are launched into space. Earthlings are thankfully shielded from most cosmic rays by the atmosphere, but airless bodies offer no protection. Only the upper face of a rock sitting on the surface of an asteroid is affected by cosmic rays (this is called 2π exposure geometry), whereas irradiation in space affects all sides of the rock uniformly (4π exposure). From the nonuniformity of exposure, it is usually possible to recognize those meteorites that were irradiated on their parent bodies before ejection into space.

Cosmic-ray exposure ages have now been measured for a large number of meteorites. These measured values agree more or less with the expected orbital lifetimes for Earth-crossing objects estimated from numerical simulations (generally less than 100 million years for most meteorites). One of the interesting observations from these data is that exposure ages for some meteorite groups tend to cluster at preferred values. For example, H-group ordinary chondrites have a distinct exposure age peak at approximately 8 million years, as shown in Figure 8.7. Such clustering suggests that a single collision broke up a large mass to produce all the meteorites in this cluster. In contrast to the situation for H chondrites, exposure ages for L-group chondrites are spread over approximately a 50 million-year range, as also shown in Figure 8.7. Recall that gas-retention ages suggest that the L-chondrite parent body was catastrophically disrupted approximately 340 million years ago. Meter-sized chunks of these meteorites were produced by successions of smaller impacts into large fragments from this riven body at later times.

Several investigators have argued that meteoroids occur as **meteor streams**, essentially collections of small rocks traveling together along

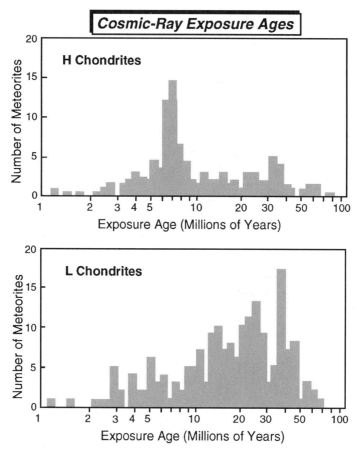

Figure 8.7: Cosmic-ray exposure ages for groups of ordinary chondrites reveal different patterns. The large peak for H chondrites suggests that these meteorites were produced by a single collision that fragmented a large mass approximately 8 million years ago. The L chondrites show a more even distribution, implying that they were chipped off larger objects by numerous small impacts during the past 50 million years or so. These diagrams are based on data reported by Kurt Marti and T. Graf (University of California, San Diego).

the same orbits. A suggestion that the H chondrites found in Antarctica represent a distinct population (that is, samples of a different meteor stream) from modern H-chondrite falls is particularly controversial. Antarctic meteorites have generally been on Earth for hundreds of thousands of years, so this difference would imply that there are temporal changes in the population of objects hitting the Earth. While such changes must certainly occur, it is not clear that meteor streams can persist for hundreds of thousands of years. In any case, the cosmic-ray exposure ages of H chondrites from Antarctica and other regions of the globe are virtually indistinguishable, casting doubt on the idea that these populations are different.

Iron meteorites have much longer exposure ages, commonly hundreds of millions of years, than their stony counterparts. Why would irons exist as small pieces in space so much longer than stony meteorites? The difference in exposure history probably simply reflects the greater strength of iron meteorites relative to rocky bodies. Irons could presumably survive small impacts that would crush stones into dust. This difference in destructability may explain why meter-sized fragments of irons can apparently persist in space for as long as a billion years in some cases, but only recently broken small pieces of stones complete the trip to Earth.

The stony meteorite with the longest cosmic-ray exposure age is the Norton County (Kansas) aubrite at 104 million years. Although the exposure ages of other aubrites are somewhat younger, they are still old compared with chondrites and other achondrite classes. If the shorter cosmic-ray exposure ages of stony meteorites are due to the weakness of these objects that limits their lifetimes in space, some other factor must protect the aubrites from collisional destruction. It has been suggested that derivation from a long-lived source body, perhaps the near-Earth asteroid 3103 that is spectrally similar to aubrites, can explain their longevity. The orbit of this particular asteroid has a high inclination, so it may experience less-frequent collisions with other bodies that lie mostly within the ecliptic plane. Impact-derived fragments from this asteroid would presumably inherit the same highly inclined orbit and thus last longer in space.

The exposure ages of lunar and Martian meteorites are even shorter than those of chondrites. Lunar meteorites generally have ages of less than a million years (often far less), and Martian meteorites fall into groups with exposure ages of approximately 3 million and 12 million years, with a few older stones. These results confirm the conclusions of numerical simulations that rocks from the Moon and Mars are ejected as small objects that make the journey to Earth within relatively short periods.

Another effect of the interaction of cosmic rays with meteoroids in space is the production of thermoluminescence (TL), which basically is energy stored in crystals of certain minerals by ionizing radiation. During irradiation in space, electrons are forced into traps (probably some sort of crystal defects), and as the meteorite is heated in a laboratory on Earth light is emitted (hence it luminesces) as the electrons are freed. Over millions of years of exposure, the traps become filled and the TL is said to have reached an equilibrium state. If the meteoroid is heated in space during a close approach to the Sun, TL occurs, just as it does during laboratory heating. Thus the measurement of the amount of TL remaining can be related to the meteorite's orbit, specifically its perihelion (minimum distance from the Sun). Most meteorites appear to have had perihelia close to 1 AU.

At the Finish Line

A runner's heartbeat and respiration rate continue at accelerated levels even after the race is finished, but will gradually slow to normal levels after a few minutes at rest. Meteorites likewise take time to recover from their interplanetary exertions, at least in terms of the decay times of the new radionuclides produced by cosmic ray exposure. This can be a useful tool to discover how long meteorite finds have been on Earth. The time of Earth residence is called the **terrestrial age** of the meteorite.

Because a fallen meteorite is shielded from further exposure to cosmic rays by the Earth's atmosphere, no new isotopes will be made by spallation, only by the decay of the radionuclides already present. If the amount of a certain isotope produced by cosmic irradiation is measured in a recently fallen meteorite, we would expect that all the other spallation-produced nuclides should be present in amounts appropriate to the same duration of exposure. However, in many meteorite finds, the concentrations of the shortest-lived radionuclides are found to be significantly less than predicted. The reason for this discrepancy is that the short-lived isotopes have decayed during the meteorite's residence on Earth, but the longer-lived radioactive isotopes produced by spallation are still present at or near their original abundances. Figure 8.8 illustrates this effect for old chondrites found in Antarctica versus more recent chondrite finds from other locations. The amounts of ^{53}Mn, a radioactive isotope of manganese produced by spallation reactions in space, are approximately the same in both populations. However, the amounts of ^{26}Al, a radioactive isotope of aluminum that decays more rapidly than ^{53}Mn, are generally lower in Antarctic chondrites.

From this simple comparison we can deduce that Antarctic meteorites, as a group, must have been frozen in the ice sheet for longer times than the terrestrial residence of non-Antarctic finds. This is one of the important features of Antarctic meteorites – they represent a population of objects

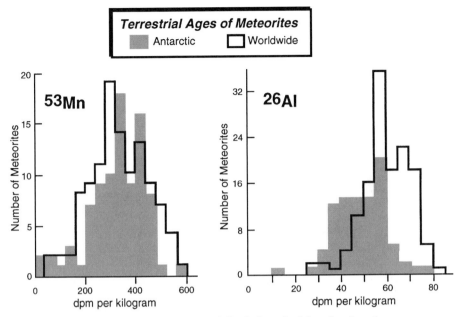

Figure 8.8: The terrestrial ages of meteorite falls, the length of time they have been on Earth, can be estimated from the relative decays of several radionuclides with different half-lives that were produced by cosmic-ray exposure. These two diagrams compare the radioactivities produced by ^{53}Mn and ^{26}Al (both expressed in disintegrations per minute, or dpm, per kilogram of sample). Data for chondrites found in Antarctica and elsewhere are compared. The amounts of radioactive ^{53}Mn are roughly the same in both Antarctic and non-Antarctic finds, but the amount of live ^{26}Al is lower in Antarctic meteorites. ^{26}Al decays more rapidly than ^{53}Mn, so this difference reflects the longer terrestrial residence of meteorites in the Antarctic environment. These data were obtained by John Evans and co-workers (Battelle Pacific Northwest Laboratories).

that orbited in space at earlier times, prehistoric in human terms. By use of several different spallation-produced isotopes that decay at different rates (Figure 8.9), it is possible to quantify terrestrial ages for individual meteorites. For example, the calculated terrestrial ages for finds from different regions of Antarctica appear to be distinct: approximately 150 thousand years for meteorites from the Allan Hills region, and approximately 400 thousand years for specimens from the area around Lewis Cliff. This variation is probably due to ice movements and different ablation rates, because all the meteorites at one locality were certainly not parts of a single gigantic fall.

Another indication of terrestrial ages is based on TL in meteorites. TL is acquired during irradiation in space, but because it gradually fades away

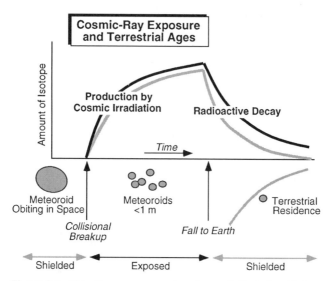

Figure 8.9: This diagram summarizes the evolution of spallation-produced radionuclides in meteorites. Most of a meteoroid is effectively shielded from cosmic rays
in space, but irradiation occurs once the object is broken by impact into fragments
of meter size or smaller. Eventually the abundance of some isotopes could reach a
plateau, representing a balance between production rate of new isotopes and their
decay. Measurement of the abundances of various isotopes yields the cosmic-ray exposure age. On falling to Earth, the meteorite is shielded from further irradiation, and
the cosmic-ray-produced isotopes begin to decay at various rates. The measurement
of two or more of these isotopes can then reveal the meteorite's terrestrial age.

once the meteorite falls to Earth, its measurement provides a relative scale
for terrestrial age.

Meteorite Journeys

The space odysseys pieced together for meteorites are based on impact
ages that may date the times these objects were ejected from their parent
bodies, numerical simulations of their orbital evolution through resonances and into the Earth's neighborhood, cosmic-ray exposure ages that
quantify their travel times as small bodies in space, and terrestrial ages
that reveal when their journeys ended. I suppose that these histories are
little more than cartoons (see Figure 8.10), but they represent a prodigious
amount of research effort expended to reach this level of understanding.
The paths by which meteorites have reached the Earth are exceedingly
complicated, but theory, observation, and measurement have converged
to provide an intriguing picture of how celestial mechanics delivers meteorites to our planet.

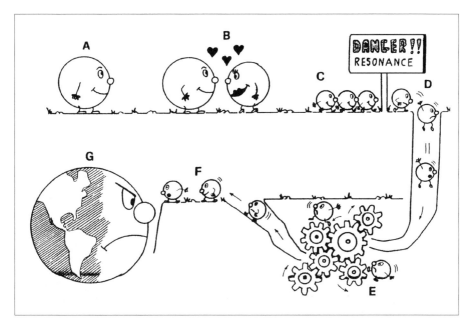

Figure 8.10: A cartoon summarizing the mechanisms by which asteroidal materials arrive on Earth. Asteroidal collisions produce fragments that are injected into resonances, which provide a complex, dynamic mechanism to increase the orbital eccentricity until the fragments arrive in the Earth's neighborhood. Courtesy of Vincenzo Zappala (Astronomical Observatory of Torino, Italy).

Suggested Readings

The topics in this chapter have not yet been treated in many nontechnical publications, and the mathematical nature of the subject makes for difficult reading. However, the references below contain a wealth of information for the serious reader.

GENERAL

Wasson J. T. (1985). *Meteorites: Their Record of Early Solar System History*, Freeman, New York.

Chapter III of this book describes cosmic-ray exposure and terrestrial ages, and Appendix H serves as a nice introduction to celestial mechanics.

EJECTION FROM ASTEROIDS

Keil K., Haack H., and Scott E. R. D. (1994). Catastrophic fragmentation of asteroids: evidence from meteorites. Planet. Space Sci. **42**, 1109–1122.

A discussion of the evidence that impact was a pervasive process affecting asteroids.

Bogard D. D. (1995). Impact ages of meteorites: a synthesis. Meteoritics Planet. Sci. **30**, 244–268.

The timing of impact events can be determined from ^{40}Ar measurements; this paper summarizes a great deal of complex literature on gas-retention ages in an easily understood manner.

Alexeev V. A. (1998). Parent bodies of L and H chondrites: times of catastrophic events. Meteoritics Planet. Sci. **33**, 145–152.

A new revision of gas-retention ages for chondrites based on ^{4}He.

EJECTION FROM PLANETS

Melosh H. J. (1984). Impact ejection, spallation, and the origin of meteorites. Icarus **59**, 234–260.

This technical paper presents calculations showing how large impacts can remove rocks from the gravitational grasp of planets.

ORBITAL EVOLUTION

Binzel R. P. and Xu S. (1993). Chips off of asteroid 4 Vesta: evidence for the parent body of basaltic achondrite meteorites. Science **260**, 186–191.

The discovery of orbiting fragments of Vesta bridging the gap between Vesta and the 3/1 mean-motion resonance.

Farinella P., Gonczi R., Froeschle Ch., and Froeschle C. (1993). The injection of asteroid fragments into resonances. Icarus **101**, 174–187.

A demonstration of how asteroidal fragments can be thrown into nearby resonances and thereby escape the asteroid belt.

Gladman B. J., Burns J. A., Duncan M., Lee P., and Levison H. F. (1996). The exchange of impact ejecta between terrestrial planets. Science **271**, 1387–1392.

A numerical simulation of the orbital dynamics of lunar and Martian meteorites.

Gladman B. J., Migliorini F., Morbidelli A., Zappala V., Michel P., Cellino A., Froeschle C., Levison H. F., Bailey M., and Duncan M. (1997). Dynamical lifetimes of objects injected into asteroid belt resonances. Science **277**, 197–201.

A numerical simulation of the orbital fate of asteroid fragments injected into mean-motion and secular resonances.

COSMIC-RAY EXPOSURE AND TERRESTRIAL AGES

Crabb J. and Schultz L. (1981). Cosmic ray exposure ages of ordinary chondrites and their significance for parent body stratigraphy. Geochim. Cosmochim. Acta **45**, 2151–2160.

A technical paper that thoughtfully summarizes a great deal of data on the cosmic-ray exposure of chondrites in space.

Evans J., Wacker J., and Reeves J. (1992). Terrestrial ages of Victoria Land meteorites derived from cosmic-ray-produced radionuclides. Smithsonian Contrib. Earth Sci. **30**, 45–56.

A tabulation of many terrestrial ages for Antarctic meteorites.

Benoit P. H. and Sears D. W. G. (1997). The orbits of meteorites from natural thermoluminescence. Icarus **125**, 281–287.

An especially clear explanation of the phenomenon of TL by its leading practictioners.

The Importance of Meteorites: Some Examples

stronomers have a unique advantage over the practitioners of other branches of science: they can literally look into the past. Astronomical distances are so vast that light from other suns often takes many millions of years to reach our planet. Thus the objects and events seen through telescopes are long past, allowing a record of progressive stellar evolution to be constructed from direct observation. It is difficult to test models that explain how stars evolve, however, because so much of what happens inside a star is not manifested in its outward appearance. Understanding the complex processes that occur within the giant molecular clouds that serve as nurseries for clusters of incubating stars is likewise challenging, because these dark blotches give off few observable signals (see Figure 9.1).

The formation of solid matter, ranging in size from dust motes to planets, around stars is actually of little consequence in terms of mass, but it is of tremendous importance from the vantage point of the beings that live on at least one such planet. Although hundreds of dusty nebulae around fledgling stars have been imaged by the *Hubble Space Telescope* and a handful of planets have been discovered outside our own solar system, astronomy still offers relatively little concrete information about the origin of worlds like the Earth or of the spawning of life that makes our own world especially interesting. Meteorites offer some tantalizing clues for understanding planet-forming and life-enabling events and processes and even insights into the hidden workings of stars and molecular clouds that existed long before the birth of the solar system.

This chapter focuses on meteorites as history. We are only beginning to decode the encrypted historical records contained in these chunks of rock and metal, but the views they have afforded so far are breathtaking.

How Elements and Molecules are Made

Before we consider what meteorites tell us about the history of the cosmos, it may be useful to review current ideas about how and where atoms

Figure 9.1: This *Hubble Space Telescope* image of the Eagle Nebula shows a dusty molecular cloud containing numerous infant stars, each embedded into the top of a fingerlike protrusion. The embryonic stars shield the gas and dust behind them from erosion by ultraviolet light from other nearby hot stars, creating the fingers. Photograph courtesy of NASA.

are created and processed into molecules. The formation of elements, a process called **nucleosynthesis**, was described in one remarkable paper by Margaret and Geoffrey Burbidge, William Fowler, and Fred Hoyle in 1957 (this paper is so influential and widely recognized that it is commonly known simply as "B^{2}FH," for the initials of the authors' surnames). B^{2}FH argued that fusion of nuclei within a star converts lighter elements into heavier ones, that is, elements with higher atomic numbers. Nuclear fusion is, of course, the source of the star's luminosity. During much of a star's lifetime (the so-called **main-sequence** stage), it burns hydrogen to make helium. Once the hydrogen fuel in the stellar core is exhausted, it may swell to become a red giant, and helium ignites to form carbon and oxygen. Stars larger than our Sun continue the progressive elemental transmutation to heavier elements, as the ashes from one fusion cycle become the fuel for the next. Because the temperature is higher in the stellar core than at the surface, the star becomes layered like an onion, with each layer the locus of a particular fusion reaction.

Nucleosynthesis actually produces specific isotopes, sometimes called **nuclides**. Fusion inside main-sequence stars proceeds at a relatively leasurely pace, so this kind of nucleosynthesis is called the **s** (for slow) **process**. The upper part of Figure 9.2, a plot of the number of neutrons versus

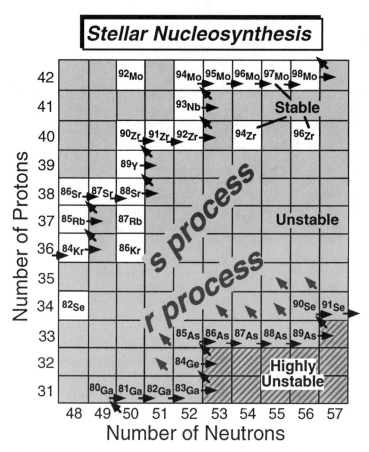

Figure 9.2: A portion of the nuclide chart illustrates how the s process (upper arrows) and the r process (lower arrows) produce new elements by addition of neutrons. The white boxes represent stable nuclides, whereas the gray boxes represent unstable nuclides that undergo radioactive decay, which shifts them up and to the left. Crosshatched boxes represent extremely unstable nuclides that decay more rapidly than neutrons can be added to their nuclei during the r process.

protons, illustrates how the s process works. Neutrons are added to a stable nuclide (represented by white boxes), pushing the resulting nuclei to the right (shown by horizontal arrows) in the diagram. When an unstable nuclide (represented by a gray box) results, it then decays by loss of alpha particles, causing it to move up and to the left (diagonal arrows) until it becomes a stable nuclide. The combined processes of neutron addition and radioactive decay continue in an orderly, stairstep fashion, creating progressively heavier elements. As temperatures inside the star eventually increase, collisions between nuclei become so violent that as many atoms are destroyed as are created. A kind of equilibrium is finally reached, in which only the most stable nuclides survive. Iron and its neighbors in the Periodic Table are very stable, and thus are the heaviest elements that stars can fabricate under s-process conditions.

Eventually red giants exhaust their nuclear fuel and collapse. Matter falling onto the compressed stellar core rebounds, causing a shock wave that reverberates outward and causes the star to explode, producing a supernova (see Figure 9.3). For a few short seconds during the supernova event, the interior of the star is raked with rapidly moving neutrons,

Figure 9.3: Two superimposed telescopic photographs show a negative (dark) image of stars in the Magellanic Cloud and a positive (bright) image of this region following the 1987 supernova. The black dot in the center of the expanding cloud is the star that exploded. Supernova explosions are the sites of r-process nucleosynthesis. Photograph courtesy of the Anglo–Australian Telescope.

causing elements heavier than iron to be produced by the so-called **r** (for rapid) **process**. The lower part of Figure 9.2 illustrates r-process nucleosynthesis. Neutrons are added much faster than the resulting unstable nuclides can decay, until finally an extremely unstable nuclide (represented by a crosshatched box) is reached. Radioactive decay then moves the nuclide up and to the left, where further neutron addition can occur, ultimately causing nuclides to march around the highly unstable region. Once the supernova event is over, the unstable nuclides decay more slowly into stable nuclides, but in this way some isotopes form that cannot be reached by s-process nucleosynthesis.

The cosmic abundance of the elements provides some confirmation of the B²FH nucleosynthetic mechanisms. The cosmic abundance table is actually constructed from analyses of chondritic meteorites (especially CI carbonaceous chondrites). Except for the most volatile elements, meteorite elemental abundances closely match solar abundances, but elements in meteorites can be analyzed with greater accuracy than those in the Sun. The atomic abundances of the various elements, relative to a million atoms of silicon, are illustrated in Figure 9.4. The sawtooth pattern results from differences in the stability of nuclides with odd and even atomic numbers (numbers of protons), the even-numbered nuclides being much more stable. The pronounced peak in abundance at iron reflects the equilibrium obtained in highly evolved red giant stars, and the downward slant of the distribution with increasing mass number (numbers of protons plus neutrons) reflects the increasing difficulty in synthesizing heavier elements in supernovae. The low abundances of lithium, beryllium, and boron are due to reactions in main-sequence stars that systematically destroy these elements. Although the cosmic abundance pattern can be rationalized by comparison with the B²FH theory of nucleosynthesis, this pattern is actually a mixture of atoms formed in many stars over many billions of years. Testing theories of nucleosynthesis requires the analysis of materials produced in individual stars or even in individual layers within a star or a supernova.

Once ejected into the interstellar medium, most elements condense as solid mineral grains, probably following a condensation sequence like that outlined in Chapter 2. Some of the carbon, hydrogen, oxygen, and nitrogen atoms are processed into organic compounds within cold, dense molecular clouds. At these temperatures, most gaseous atoms freeze out onto the surfaces of solid grains. The condensation process generates new molecules, and exposure of these icy mantles to ionizing radiation can create other, more complex compounds. Gaseous organic molecules in interstellar clouds have been identified by radio astronomers, and the current list contains more than a hundred molecules ranging in size from

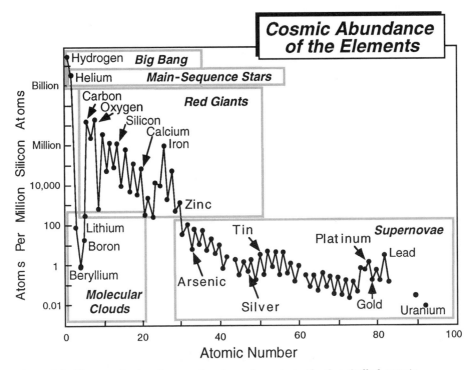

Figure 9.4: The atomic abundances of various elements in the Sun (called cosmic abundances), each relative to one million silicon atoms. The elements are ordered from left to right according to increasing atomic number (the number of protons they contain). Each marked division on the vertical scale signifies a tenfold increase in abundance. Cosmic abundances for most elements are obtained from analyses of chondritic meteorites. Each interior box encloses elements that formed in a similar astrophysical setting.

two to thirteen atoms. The identification of organic molecules in solid form within these clouds is more difficult, but infrared spectroscopy can discern their presence. Prominent absorption features are due to stretching of carbon–hydrogen bonds in complex compounds such as **polycyclic aromatic hydrocarbons** (fused hydrocarbon rings, commonly referred to as PAHs).

The incorporation of presolar mineral grains and organic matter into the solar nebula affords other opportunities for evolution. Heating of infalling grains as they pass through a shock front defining the edge of the nebula may cause evaporation and **pyrolysis** (organic reactions resulting from loss of water). Eventually organic molecules become sequestered with rock and ices to form planetesimals or comet nuclei. Heating of rocky bodies causes ices to melt, and the combination of water and higher temperatures produces mixtures of larger, more complex organic molecules (**kerogens**) from simpler precursor molecules. Even comets that have escaped signficant heating contain an array of organic molecules, and kerogenlike mixtures have been isolated from some IDPs thought to have been derived from comets.

Matter Older than the Solar System Itself

If chondrites are primitive materials surviving from the time of the solar system's birth, it is possible that they might still contain some presolar materials. The first hint that interstellar matter might be present in chondrites came in 1964, when researchers found xenon with an unusual isotopic composition (now called Xe–HL, signifying that it is enriched in both the heaviest and the lightest isotopes). This exotic xenon component is so different from normal planetary xenon that it cannot be inherently part of our solar system, and its release from chondrites over a narrow temperature interval implies that Xe–HL resides in some specific mineral. More than twenty years passed before Edward Anders and his colleagues at the University of Chicago managed to separate the mineral carrier by using Xe–HL to trace its location as the host meteorite was progressively dissolved in a series of acids. After the chemical destruction of more than 99.9% of the meteorite, all that was left was a fine white powder, consisting of miniature diamond crystals (see Figure 9.5). These

Figure 9.5: These diamond crystals are interstellar grains, formed as carbon-rich vapor released from red giant stars cooled and condensed. The stardust grains are extremely small (viewed here with an electron microscope), averaging only approximately 25 Å across. The grains are laced with an exotic mixture of noble-gas isotopes that were generated during a nearby supernova event and implanted into the diamonds. Photograph by Roy Lewis (University of Chicago).

microdiamonds are so small (approximately 25 Å on average) that their properties are dominated by their large surface area. They are apparently the most abundant **interstellar grains** in chondrites, with concentrations of several hundred parts per million. These diamonds probably condensed as the carbon-rich outer layers of distended red giant stars were sloughed off and cooled. The distinctive Xe–HL was produced during a supernova, so this material is thought to have been ejected in an energetic stellar wind that overtook clouds of diamond stardust and was implanted into the tiny grains.

In addition to diamonds, several other kinds of interstellar grains have now been isolated from chondrites. The present list includes silicon carbide (the synthetic form of this mineral, used as an industrial abbrasive, is known as carborundum), graphite (like diamond, a form of carbon), corundum (aluminum oxide), and silicon nitride. Electron microscopes have also revealed minute inclusions of other minerals within some silicon carbide and graphite grains, identified as carbides of titanium, zirconium, and molybdenum (see Figure 9.6).

The ability to analyze interstellar grains in the laboratory means that a wide range of elements can be measured in them with high precision. For example, silicon carbide grains have been isotopically analyzed for carbon, nitrogen, silicon, magnesium, calcium, titanium, strontium, barium, samarium, neodymium, helium, neon, argon, xenon, and krypton. Such data allow inferences to be drawn about the origin of these grains. Isotopic anomalies, such as excess amounts of ^{44}Ca, the calcium decay

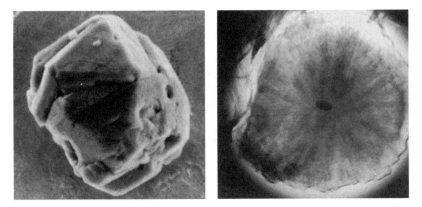

Figure 9.6: Electron microscope images of other kinds of interstellar grains. The left photograph shows a silicon carbide crystal isolated from the Murchison (Australia) carbonaceous chondrite. The right image reveals a minute inclusion of titanium carbide (black) inside a spherule of interstellar graphite. Each grain is approximately 5 μm in diameter. Photographs by Sachiko Amari and Tom Bernatowicz (Washington University).

product of the short-lived titanium radioisotope ^{44}Ti, clearly indicate formation by r-process nucleosynthesis during supernova explosions. Other interstellar grains have compositions consistent with s-process reactions that occur within red giant stars. As one illustration, let us examine the isotopic differences between grains of silicon carbide that are thought to have formed in the evelopes of red giant stars and those that formed during supernova events. Figure 9.7 demonstrates that the two populations of silicon carbide grains have distinctive isotopic compositions of silicon, carbon, and nitrogen. Whether they formed in stars or supernovae, the mixtures of isotopes in interstellar grains provide critical tests for models of stellar nucleosynthesis.

Another possible type of interstellar grain is found as inclusions within tiny IDPs that were probably derived from comets. These minute inclusions consist of amorphous (having no crystal structure, like a glass) silicate material containing bits of iron–nickel metal and iron sulfide, giving rise to the acronym **GEMS** (glass with embedded metal and sulfides). An important characteristic of GEMS that supports their interstellar origin is the severe irradiation they have experienced. The interstellar medium is a hostile environment for small grains because of cosmic-ray bombardment, and telescopic observations suggest that silicate grains in the interstellar medium may be mostly amorphous (like glass) because of radiation damage.

Chondritic meteorites also contain organic matter, some of which could have been inherited directly from interstellar space. Reactions between ions and molecules that occur in frigid molecular clouds favor the incorporation of deuterium (the heavy isotope of hydrogen) into organic compounds. Isotopic measurements of hydrogen in the organic matter of carbonaceous chondrites indicate that these compounds are markedly enriched in deuterium over terrestrial organic materials (see Figure 9.8), suggesting that they were derived from matter in interstellar space. However, the organic compounds in chondrites are typically complexly branched hydrocarbons, probably crafted in asteroids from the deuterium-enriched materials originally made in molecular clouds.

Nevertheless, it may be that at least some of the organic matter in meteorites was inherited intact from the cosmos. This inference comes from detailed examination of the structures of organic molecules. Many organic compounds, such as amino acids that are necessary ingredients for all terrestrial life, come in two versions, basically mirror images of each other (see Figure 9.9). These **stereoisomers** are identified as left and right handed, based on their ability to rotate light in one direction or another.

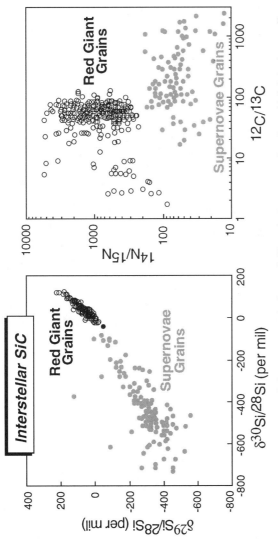
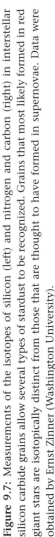

Figure 9.7: Measurements of the isotopes of silicon (left) and nitrogen and carbon (right) in interstellar silicon carbide grains allow several types of stardust to be recognized. Grains that most likely formed in red giant stars are isotopically distinct from those that are thought to have formed in supernovae. Data were obtained by Ernst Zinner (Washington University).

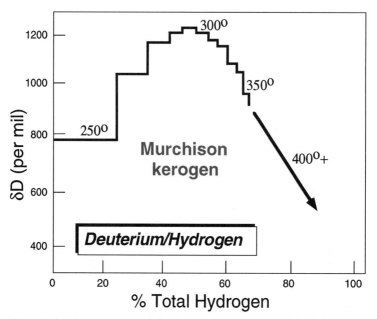

Figure 9.8: Measurements of the isotopic composition of hydrogen (expressed as the ratio of deuterium to normal hydrogen, relative to a standard) in the Murchison carbonaceous chondrite. Stepwise heating experiments release hydrogen from various materials at different times; steps between 250 and 350 °C represent hydrogen released from organic matter. The extreme deuterium enrichment suggests that these compounds are related to hydrocarbons formed in interstellar space. Data were obtained by John Kerridge and colleagues (University of California at Los Angeles).

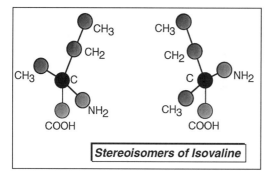

Figure 9.9: Organic molecules can occur in left- and right-handed forms (stereoisomers), depending on the geometric attachment of various hydrocarbon groups. These sketches compare the isomers of isovaline, an amino acid found in carbonaceous chondrites. The left-handed isomer of this compound is slightly more abundant in the Murchison chondrite.

When created in the laboratory, these compounds are invariably **racemic mixtures**, that is, they contain equal numbers of left- and right-handed molecules. However, all living organisms on Earth use only left-handed amino acids, apparently a trait inherited from a common ancestor. For many years, chemists have thought that the handedness of the organic molecules formed by living organisms was a peculiar characteristic, even a signature, of life. To the extent that terrestrial life uses one stereoisomer exclusively, that is probably true. However, new analyses of the amino acids in carbonaceous chondrites by John Cronin and Sandra Pizzarello at Arizona State University indicate a small preference for left-handed molecules, amounting to only a few percent above the equal ratios of racemic mixtures. The amino acids that were analyzed are unknown in the Earth's biosphere, so this peculair excess may imply a slight preference for left-handed isomers in the interstellar medium. Cronin and Pizzarello speculate that organic molecules formed in interstellar space might result from exposure to circularly polarized light emitted from neutron stars. If this is correct, it would imply that these organic compounds were inherited directly from molecular clouds, without processing in meteorite parent bodies. These molecules would thus qualify as intact interstellar matter.

Materials Formed in the Solar Nebula

The solar nebula takes its name from the Latin word for fuzzy, an allusion to its probable appearance (see Figure 9.10). Rotation of the gas and the dust that eventually coagulated to form the solar system would have caused the nebula to flatten into a disk. The *Hubble Space Telescope* has now imaged hundreds of disks, each containing an infant protostar, providing confirmation of the idea that our own solar system began in this manner. Most of the constituents of chondrites formed within the solar nebula and thus provide important information on nebular processes and conditions.

We have already discussed the kinds of chemical separations (fractionations) that occurred in the early solar system, as revealed by chemical analyses of chondrites. These include fractionation of volatile elements from refractory elements, accomplished during condensation of solids from nebular gases. It is not clear, however, that condensation was a pervasive, nebulawide process. Instead, many chondrite components seem to have experienced both condensation and evaporation. For example, one class of refractory inclusions in carbonaceous chondrites contains

Figure 9.10: The solar nebula was a disk in which planetesimals accreted from gas and dust, and planets accreted from planetesimals. The fledgling star in the center is erupting in violent outbursts, ejecting matter from its poles out of the plane of the disk. Painting by William Hartmann (Planetary Science Institute).

depletions of both the most refractory and the most volatile rare-earth elements, which might be explained by partial condensation following an earlier episode of partial evaporation. The cooling history of the nebula was undoubtedly complex and may have involved cyclic temperature changes.

Fractionations between solids also occurred in the nebula. We previously learned that the various chondrite groups show differences in their proportions of metal and silicate. Another kind of fractionation pattern between solids is illustrated in Figure 9.11, a plot of silicon/aluminum versus magnesium/aluminum. Aluminum is refractory, magnesium is moderately refractory, and silicon is not refractory. Two distinct trends

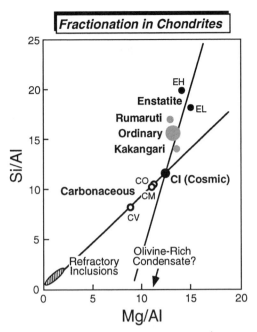

Figure 9.11: Fractionations between solid materials in the solar nebula are revealed by systematic variations among chondrite groups. Silicon (Si), aluminum (Al), and magnesium (Mg) all formed solid silicate grains in the nebula. Refractory inclusions contain minerals rich in aluminum, whereas olivine grains are rich in magnesium. Starting with a cosmic (CI chondrite) composition, the compositions of other carbonaceous chondrites can be derived by addition of refractory inclusions. The compositions of other chondrite groups might be related to the cosmic composition by loss of an olivine-rich condensate.

among chondrite groups are apparent, intersecting at the cosmic abundance composition (taken as the starting point). The compositions of the various carbonaceous chondrite groups reflect the addition of refractory inclusions, whereas the other chondrite groups require loss of a component with approximately the composition of olivine.

Sorting of the various components of chondrites, such as chondrules, refractory inclusions, metal and sulfide grains, and matrix dust, may reveal something about the nature of the nebula itself. Chondrules in any chondrite have a restricted range of sizes, but different chondrite groups have distinct chondrule-sized populations. Metal grains, too, show size sorting, but they tend to be smaller than the chondrules in the same meteorite. This suggests that chondrules and metal grains may have been sorted together by mass. Although not strictly mass equivalent, these objects are similar enough that they could have been aerodynamically sorted within turbulent eddies in the nebula.

The physical conditions within parts of the nebula can also be specified from studies of chondrite components. The temperatures necessary to melt chondrules and refractory inclusions are in the range of 1,500 to 1,700 °C. Rapid cooling from peak temperatures suggests transient heating events on local rather than nebulawide scales. The presence of chondrules within chondrules indicates multiple periods of heating and cooling. Compared with chondrules, refractory inclusions were heated for longer periods, accounting for their greater fractionation of volatile elements, and were cooled more slowly. No completely satisfactory mechanism for rapid heating and cooling of nebula materials has been identified. The abundances of volatile trace elements and water in bulk chondrites require that nebula temperatures had fallen to several hundred degrees Celsius or lower before the onset of accretion. Pressures in the nebula are thought to have been in the range of 1 to 100 mTorr based on condensation calculations, but these estimates are not definitive.

The oxidation state of the nebula must have varied considerably. Enstatite chondrites are highly reduced, whereas carbonaceous chondrites are highly oxidized. Neither of the redox conditions implied by the mineralogy of these chondrite groups is consistent with that predicted for a gas of solar composition. Variations could have resulted from differences in the ratio of gas (rich in hydrogen, which renders the environment reducing) and dust (rich in oxygen, which makes the environment more oxidizing).

Frank Shu and his colleagues of the University of California, Berkeley, have attempted to reconcile the various properties of chondritic materials with astrophysical models of the solar nebula. Their conclusions draw on observations of the violent behavior of newborn (T Tauri) stars. These stars radiate energy at a furious pace and sometimes exhibit wild oscillations in brightness, possibly powered by accretion of matter from the surrounding nebula in fits and starts. Ejected materials emanate from the stars' rotational poles, producing bipolar outflows. Shu and his colleagues suggested that refractory inclusions and chondrules might have been formed when small balls of accreted dust were lifted by solar winds out of the relative cool of a shaded disk close to the star and into the heat of direct sunlight. This mechanism also offers a means of sorting to explain the size distribution of these objects. After being launched out of the ecliptic plane in biopolar outflows, the chondrules and the inclusions would fall back into the disk at great distance from the Sun, where they would accrete with dust to form chondrites.

The duration of the solar nebula can be estimated by the study of high-resolution isotopic clocks in meteorites. The time of formation for refractory inclusions has been fixed rather precisely at 4.566 billion years, by use of the lead isotopes formed by radioactive decay of uranium. These are the oldest known objects in the solar system, and they contain the decay products of now-extinct nuclides like ^{26}Al that would only have been incorporated live if these objects formed soon after the solar nebula appeared. HED achondrites have ages that are only a few million years younger, implying that Vesta had already formed and differentiated in that interval. Thus dust and gas in the solar nebula probably condensed and accreted to form planetesimals within a few million years. Because accretion was so fast, heating of asteroid-sized objects by decay of ^{26}Al offers a plausible explanation for the thermal processing experienced by so many meteorite parent bodies.

Formation and Differentiation of Planets

The formation of the planets is thought to have occurred by accretion of small planetesimals, essentially asteroids. The chemical and the isotopic compositions of meteorites thus offer an interesting comparison with planet compositions. A difficulty arises, though, because the Earth and other planets are differentiated; consequently, it is impossible to find any rock on or within such a body that has the composition of the whole planet. We can circumvent this problem by considering ratios of elements that are geochemically similar, that is, elements that are sequestered together during planetary differentiation. Although the abundances of various elements may be altered during melting and crystallization, the ratios of two elements that have similar geochemical behaviors should not be.

The terrestrial planets contain refractory elements in approximately chondritic proportions, as shown in the left-hand side of Figure 9.12. Lanthanum and uranium are both highly refractory elements with similar geochemical behaviors. The ratio of these elements is uniform in rocks from the Earth, Moon, Mars, and asteroid Vesta, having a value that is about the same as the chondritic ratio. However, planets are strongly depleted in volatile elements. The right-hand side of Figure 9.12 shows a plot of potassium, a volatile element, against refractory lanthanum. These two elements have similar geochemical behavior, so planetary differentiation does not significantly alter the ratio of the two elements; thus samples from each planet define a line on the diagram. All the planets are depleted in potassium relative to lanthanum, but to varying degrees.

Element Fractionation

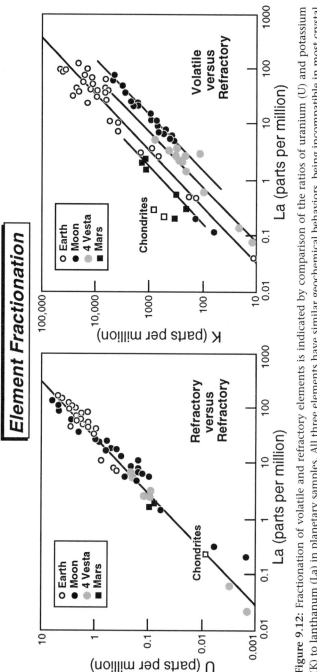

Figure 9.12: Fractionation of volatile and refractory elements is indicated by comparison of the ratios of uranium (U) and potassium (K) to lanthanum (La) in planetary samples. All three elements have similar geochemical behaviors, being incompatible in most crystal structures, so their ratios should not be affected by differentiation. Abundances of these elements in four solar system objects are shown; asteroid 4 Vesta is represented by HED meteorites, and Mars by SNC meteorites. Uranium and lanthanum are both refractory, and samples from all four bodies have the same uranium/lanthanum as chondrites. Potassium, however, is volatile, and its ratio to refractory lanthanum is lower in all four bodies than in chondrites. Each body seems to show a characteristic potassium/lanthanum ratio, with Mars having the least volatile depletion and the Moon having the greatest. The data were obtained by Heinrich Wänke and co-workers (Max-Planck-Institut, Mainz, Germany).

The depletion of volatile elements in planets has sometimes been as-cribed to volatile loss during planetary accretion. However, not all the atoms of volatile potassium would have been fractionated equally. Dur-ing volatilization, the light isotope ^{39}K would have been preferentially lost relative to heavy ^{41}K. Munear Humayun and Robert Clayton at the University of Chicago observed that the isotopic ratio ^{39}K/^{41}K is uniform from planet to planet, despite differences in the amounts of potassium relative to refractory elements. This is a strong argument against volatile loss during planet formation. Volatile elements were already depleted in planetary building blocks by the time they accreted to form larger bodies.

Planet compositions are sometimes modeled as mixtures of various chondrite groups or chondrite components. Although the idea that plan-ets were made from primitive chondritic meteorites is appealing, no mix-tures of the known chondrite groups can satisfy all the chemical con-straints. Figure 9.13 compares the oxygen isotopic compositions of Earth,

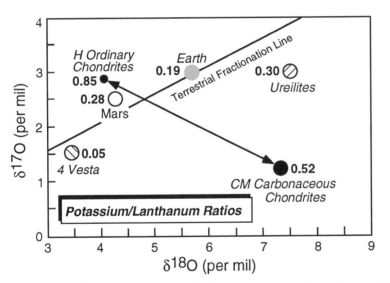

Figure 9.13: The oxygen isotopes and volatile-to-refractory-element ratios in plan-ets do not coincide with known chondrite groups, and they cannot be readily ex-plained by mixing chondritic materials. This diagram shows the oxygen isotopic compositions (expressed as ratios to ^{16}O, relative to a standard) of the Earth, Mars (SNC meteorites), and asteroid 4 Vesta (HED meteorites), as well as ordinary and carbonaceous chondrites and ureilite achondrites. The potassium/lanthanum ratios are indicated as boldfaced numbers next to the dots. No simple mixtures of mete-orites, such as H chondrite plus CM chondrite whose mixtures would lie along the arrow connecting them, can explain both the oxygen isotopic compositions and the potassium/lanthanum ratios of these solar system bodies.

Mars (SNC meteorites), and 4 Vesta (HED meteorites) with those of ordinary and carbonaceous chondrites. The numbers next to each data point are the planetary potassium/lanthanum ratios. No mixture of the meteorites shown can reproduce both the oxygen isotopic composition and potassium/lanthanum ratio of any of these bodies.

Like the planets, achondritic meteorites show depletions of volatile elements. Although it is not possible to mix known achondrite groups to obtain satisfactory matches for planet compositions, this shared property supports the idea that planets were constructed from already differentiated planetesimals. This revelation has some interesting implications. For example, the formation of the Earth's core would have been greatly facilitated by accreting bodies having large, perhaps still-molten cores. Seismic studies indicate that our planet's core contains an element lighter than iron, but currently unidentified. Oxygen has been suggested as a possibility, because its solubility in molten iron increases with pressure. If, however, the core was assembled from smaller cores formed at low pressures inside planetesimals, sulfur is a more likely candidate. Sulfur has greater solubility in iron at low pressure, and it is a common constituent of iron meteorites.

Radiogenic isotopes of certain siderophile elements provide timescales for planetary differentiation. An isotope of tungsten, ^{182}W, is the decay product of the short-lived radionuclide ^{182}Hf, an isotope of hafnium. Both elements are highly refractory, so planets should contain them in chondritic abundances. Tungsten is strongly partitioned into iron metal and thus would be drawn into the core, if differentiation had occurred after ^{182}W decayed. Hafnium, however, is not siderophile, so it would remain in the mantle during core formation. The relatively high concentrations of radiogenic ^{182}W in SNC meteorites measured by Der-Chuen Lee and Alex Halliday at the University of Michigan point to tungsten depletion in the Martian mantle during core separation within the first 30 million years of solar system history. Planetary differentiation apparently coincided with or occurred soon after accretion.

Meteorites and Water

Life, as we can envision it, must have water. Meteorites provide some interesting perspectives on how the Earth got its water. Of course, water is merely one of a number of volatile elements and molecules, and these constituents probably share a common heritage. Noble gases are particularly revealing volatiles, because they are unreactive with the crust. They are such heavy atoms that the Earth's gravitational grasp grips them tightly, minimizing their escape into space. The abundances of noble gases in the atmospheres of the planets (expressed in Figure 9.14 in terms of nonradiogenic isotopes whose abundances do not change with time) suggest that planetary volatiles were not captured directly as gases from the solar nebula. Planetary atmospheres are depleted in neon and argon, unlike the solar abundance values that must have characterized the nebula. Instead, the atmospheric values resemble those in chondritic meteorites, suggesting that volatiles may have been carried by accreting planetesimals. This seems to be inconsistent with an earlier conclusion that planets accreted

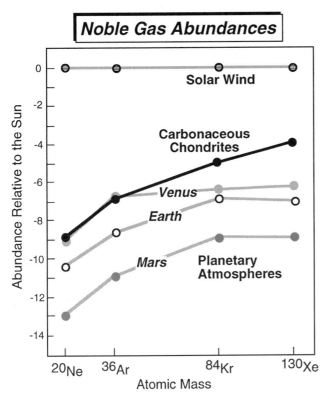

Figure 9.14: The abundances of noble gases in the atmospheres of Venus, Earth, and Mars are very different from solar abundances, but are similar to those in chondritic meteorites. This figure illustrates the measured concentrations of nonradiogenic isotopes of planetary neon (Ne), argon (Ar), krypton (Kr), and xenon (Xe), divided by those in the Sun, shown at the top. All abundances are normalized to one million atoms of silicon. The differences between the various gas reservoirs are so great that the vertical scale is logarithmic. Planetary atmospheres could not be gases snatched from the solar nebula (which would have solar composition), but may have been carried in during accretion of chondritic material.

from differentiated planetesimals that were already depleted in volatiles. A way around this problem may be to add a thin veneer of chondritic material onto planets late in their accretional histories. Carbonaceous chondrites contain as much as 20% water, so a late-accreting veneer of such material might account for some fraction of our planet's oceans as well.

Meteorites themselves provide some hints that heterogeneous accretion may have taken place. Regolith breccias from chondrite and achondrite parent bodies commonly contain clasts of altered carbonaceous chondrite, but such clasts are unknown from breccias that did not form on asteroid surfaces. Thus volatile-rich carbonaceous chondrite meteoroids apparently filtered into the inner asteroid belt after these bodies had formed and were incorporated into their surface soils. Similarly, they may have also reached the terrestrial planet region during the waning stages of accretion. The water in such meteoroids would have eventually been incorporated into the outer portions of the planets, from which it was subsequently outgassed by volcanism.

The ancient highlands of Mars were once scoured by torrents of running water, producing branching networks of valleys and huge outflow channels like that onto which the *Mars Pathfinder* spacecraft landed. Today, however, liquid water is not stable anywhere on the Martian surface, and ice is known to exist only at the poles. Consequently, it is difficult to understand how much water may have been present or what may have happened to it. SNC meteorites may provide a means of assessing the inventory, outgassing history, and evolution of Martian water. Models of the bulk composition of Mars based on these meteorites curiously indicate a planet with a low water abundance but high concentrations of other volatile elements. This might be explained by reaction of water and metallic iron in accreted materials, stripping the oxygen from water to form iron oxide with the resultant loss of hydrogen from the planet. The interior of Mars would thus be rather dry and highly oxidized. Partial melting of the Martian mantle produced the magmas that crystallized to form SNC meteorites. These magmas were rich in oxidized iron but poor in water. Consequently, the amount of water outgassed from the Martian interior over time has probably been modest. Expressed as a global ocean of uniform depth, the outgassed water on Mars may amount to no more than a few hundred meters, compared with 2.7 km for the Earth.

The small amounts of water in SNC meteorites are bound within tiny grains of hydrated minerals, such as amphibole, mica, and apatite. Laurie Leshin and her co-workers at the California Institute of Technology have measured the isotopic composition of hydrogen in these grains. They found huge enrichments in the heavy isotope (deuterium), a characteristic also shared by the Martian atmosphere. This measurement implies that the water in these rocks must have cycled through the atmosphere before being incorporated back into the crust. The meteorites provide evidence of the migration path of water that could only be guessed at from orbital photographs.

Meteorites and Life

It is increasingly apparent that some of the early Earth's **biogenic** molecules (those utilized in living organisms) were probably inherited from extraterrestrial sources. With the exception of hydrogen produced in the Big Bang, the biogenic elements (carbon, hydrogen, oxygen, nitrogen, sulfur, and phosphorus) were synthesized within stars by fusion reactions. After ejection from their stellar birthplaces, these atoms combined to form organic molecules within molecular clouds. Such molecules are not complex by the biochemical standards of living systems, but they represent a major step in organic evolution. Further thermal processing in the solar nebula and in planetesimals resulted in organic molecules of greater complexity.

The cumulative inheritance of organic processes in interstellar space, in the nebula, and in planetesimals is preserved in meteorites, especially carbonaceous chondrites. Among these molecules can be found many compounds common to terrestrial organisms, including most of the amino acids that are necessary ingredients for life as we know it. Accretion of organic matter to the early Earth, in the form of meteorites and IDPs, may have provided the raw materials for life. It is thus plausible that the evolution of life did not have to start from scratch by synthesizing organic molecules from their constituent elements.

The possibility that the Earth was salted not just with organic matter, but with life itself, by means of meteorites is an idea with a long and tortured history. In the nineteenth century, Otto Hahn described structures in iron meteorites that he inferred were the petrified remains of algae and ferns, as well as fossilized corals, sponges, and crinoids in chondrites. His imaginative interpretations were received with scathing and well-deserved criticism. In the 1930s, bacteriologist Charles Lipman cultured living cells from meteorites, and his announcement generated a response similar to that received by Hahn. When the experiments were duplicated by others, the meteoritic cells were found to be identical to common bacterial contaminants in laboratories. Apparently sterilization had not completely eradicated organisms that had worked their way into meteorites after their arrival on Earth. In the 1960s, Bartholomew Nagy and his colleagues discovered tiny clumps of matter, which they called organized elements, in meteorites and suggested that these objects were fossils. Subsequent work showed that the organized elements were oddly shaped mineral grains, starch, and ragweed pollen.

Given this track record, it is understandable that the scientific community might be skeptical of further claims for life in meteorites. In 1996, David McKay and colleagues from NASA's Johnson Space Center and Stanford University proposed that a Martian meteorite, ALH84001, contains biochemical markers, biogenic minerals, and microfossils of extraterrestrial origin. This intriguing hypothesis has engendered a great deal of public interest and scientific controversy, as would be appropriate for any discovery of such magnitude.

ALH84001 is an ultramafic igneous rock, cut by fractures that contain small carbonate grains (see Figure 9.15). The meteorite is much older (it crystallized approximately 4.5 billion years ago and was heavily shocked at 4.0 billion years) than other Martian meteorites, although the time of formation of the carbonates has not been unequivocally determined. The conditions of carbonate formation remain in dispute, with some workers favoring precipitation from fluids at low temperatures and others arguing for formation at high temperatures that are inhospitable for life (terrestrial organisms do not live at temperatures higher than approximately 120 °C). Within the carbonates are a variety of tiny grains of magnetite and sulfide, some with unusual morphologies. McKay and his co-workers suggested that these grains resembled biogenic minerals formed by some terrestrial bacteria. It is difficult to prove or refute this claim, because it is not clear to what degree size and shape are sufficient criteria for identifying biogenic minerals. Other workers have recognized elongated magnetite whiskers (see Figure 9.16) in the same carbonates, and their peculiar morphologies and growth mechanisms suggest vapor deposition, a process that has nothing to do with biology. The sulfur isotopic compositions of sulfides associated with the carbonate in ALH84001 are unlike those produced by sulfate-respiring bacteria on Earth. The presence of polycyclic aromatic hydrocarbons was also cited as supporting evidence for life. PAHs themselves serve no biologic role, but they might be produced from the decaying remains of organisms. However, most of, if not all, the organic matter in this meteorite now appears to be terrestrial contamination, a result of the meteorite's long residence in the Antarctic ice. The microfossils (more properly nanofossils) described in this meteorite (Figure 9.16) are much smaller than terrestrial bacteria, and some biologists have argued that the volume encompassed by these forms is too tiny

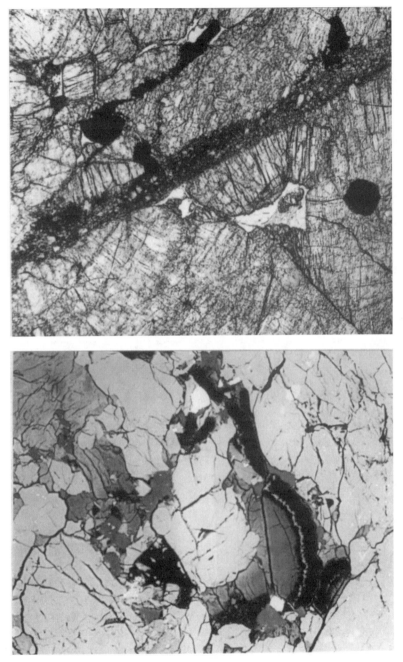

Figure 9.15: The ALH84001 Martian meteorite consists mostly of highly shocked orthopyroxene, cut by fractures, as illustrated in the upper microscopic image. The backscattered electron image (bottom) illustrates a chemically zoned carbonate grain in the fracture. Calcium-rich carbonate is gray, magnesium-rich carbonate is black, and iron-rich carbonate is white. Some scientists have suggested that these carbonates contain evidence for early Martian life. The horizontal axis of the upper figure is 0.5 cm and of the lower figure is 150 μm. Images courtesy of David Mittlefehldt (Lockheed Martin Engineering Science Services).

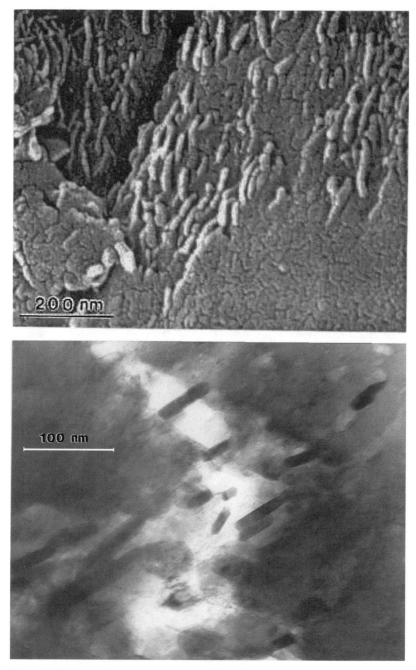

Figure 9.16: Comparison of a scanning electron microscope image of possible micro-fossils in ALH84001 carbonate (top) with a transmission electron microscope image of tiny magnetite whiskers in the carbonate (bottom). The similarity in size and mor-phology suggests that at least some of the microfossils may be magnetite, casting doubt on their validity as an indication of life. Photographs courtesy of NASA and John Bradley (MVA, Inc.), respectively.

to house enough organic matter to support the necessary biochemical activities of living cells. Conceivably, the nanofossils could represent the shrunken remains of once larger organisms. However, they are similar in size and shape to magnetite whiskers, and some workers have suggested that the apparent segmentations in some nanofossils are artifacts of the metallic coating applied during preparation of the sample for electron microscopy. The controversy over life in ALH84001 has not been fully resolved. Even if the hypothesis of Martian life is wrong, this work has demonstrated that soluble minerals, microscopic particles with unusual morphologies, isotopic anomalies, and perhaps some organic matter can survive in rocks for long periods of time, which gives hope for a Mars exploration program focused on the search for life.

The recognition that planets can swap rocks on relatively short time scales has reopened the question of **panspermia**, the idea that life could be transported from one planet to another. But if meteorites can be givers of life, they can also take it. It seems inevitable that large impact events would have biologic consequences, and sometimes impacts must have been devastating, as in the case of the mass extinction at the Cretaceous–Tertiary boundary. It is incumbent on humans to take the necessary measures to prevent such a catastrophe if it is within our means. Assembling a complete inventory of the nature and orbits of near-Earth asteroids, as well as studying how to alter an asteroid's orbit, would seem prudent.

A Final Note

The rich sound of a pealing church bell is due to its many vibrational modes, each ringing out at a different harmonic frequency. Similar modes occur in any vibrating metal plate, which develops regions that remain stationary (nodes) and regions where motion is intense (antinodes). Nearly 200 years ago, Ernst Chladni discovered that when sand was sprinkled onto vibrating plates, the different vibrational patterns were made visible as the particles danced away from the antinodes and settled on the nodes. Physical principles are rarely so graphically displayed, perhaps accounting for the popularity of this simple exercise in physics laboratories. It is this experiment for which Chladni is best known. This is the same Ernst Chladni whose pioneering work first made the study of meteorites a legitimate scientific endeavor. It is somewhat disconcerting, at least to me, that Chladni's physics experiment should have gained him more recognition than did his role as the father of meteoritics, but the importance of meteorites is not always recognized, even today.

What is the value of meteorites? The trickle of interplanetary and interstellar matter that arrives each year on Earth brings with it new and often unexpected insights into the origin and evolution of our world and its neighborhood. It is probably safe to say that Chladni would be astounded at the information that has been wrested from meteorites since he first argued for their extraterrestrial origin. And the propects for more surprises are excellent. The lifetimes for materials in Earth-crossing orbits are so short that we have seen only a tiny fraction of parent body diversity, the results of a few fairly recent collisions among asteroids or impacts into the Moon and Mars, and probably a few spent comets. Although the Antarctic meteorite collections sample populations of objects that orbited many thousands of years ago, these are still dominated by the same kinds of meteorites that fall today. A handful of unique meteorites defy classification into existing groups and must represent pieces of otherwise unsampled parent bodies. Thus the expectation that an exciting discovery will attend each new meteorite recovery is always there.

This book is an attempt to explain the scientific value of this limited meteoritic sample, as well as to describe what is known or inferred about their parent bodies. Geologic inquiry requires examination of large quantities of rock in the field and laboratory before proffering any conclusions about how any part of the Earth was formed. Imagine trying to characterize entire worlds, albeit small ones in many cases, from the study of only a few kilograms of sample and without benefit of direct field observations. It almost seems impertinent. Yet this is the task that meteoritics has set itself. It is therefore obvious why each meteorite fall or find is so important, even if it is another ordinary chondrite. We have a curious need to understand our origin and our cosmic surroundings, and meteorites provide otherwise unobtainable information to help us in this quest.

Suggested Readings

Some of the subjects discussed in this chapter have been treated in popular science articles and books. The readings below tend to be more technical, providing a more thorough background for serious readers.

INTERSTELLAR MATTER IN METEORITES

Anders E. and Zinner E. (1993). Interstellar grains in primitive meteorites: diamond, silicon carbide, and graphite. Meteoritics **28**, 490–514.

An superb summary of research on stardust, written by people who invented the subject.

Cronin J. R. and Pizzarello S. (1997). Enantiomeric excesses in meteoritic amino acids. Science **275**, 951–955.

This groundbreaking study found that some organic compounds in chondrites are not racemic mixtures and suggested that they may have formed in interstellar space.

Sandford S. A. (1996). The inventory of interstellar materials available for the formation of the solar system. Meteoritics Planet. Sci. **31**, 449–476.

An excellent review paper comparing astronomical observations of materials in the interstellar medium with laboratory studies of stardust and organic compounds in meteorites.

SOLAR NEBULA

Kerridge J. F. (1993). What can meteorites tell us about nebular conditions and processes during planetary accretion? Icarus **106**, 135–150.

A thoughtful essay describing how chondrites constrain temperature, pressure, and time, and how they reveal the details of nebular condensation, mixing, fractionation, and other processes in the solar nebula.

Shu F. H., Shang H., and Lee T. (1996). Toward an astrophysical theory of chondrites. Science **271**, 1545–1552.

An interesting hypothesis relating the properties of chondritic meteorites to what is known about young stars and the solar nebula; technically challenging reading.

PLANET FORMATION

McSween H. Y. Jr. (1989). Chondritic meteorites and the formation of planets. Am. Sci. **77**, 146–153.

This article describes, in basic terms, the formulation of models for planet chemistry based on what has been learned from chondritic meteorites.

Taylor S. R. (1992). *Solar System Evolution: A New Perspective*, Cambridge U. Press, New York.

This book contains a great deal of up-to-date information on the origin of planets. Chapters 3 and 5 are particularly relevant to topics covered in the present chapter.

Wänke H. and Dreibus G. (1988). Chemical composition and accretion history of the terrestrial planets. Philos. Trans. R. Soc. London A **325**, 545–557.

A more rigorous treatment of planetary accretion, with special attention to the idea that planets formed by mixing of several types of chondritic materials.

VOLATILE ELEMENTS AND WATER

Humayun M. and Clayton R. N. (1995). Potassium isotope cosmochemistry: genetic implications of volatile element depletion. Geochim. Cosmochim. Acta **59**, 2131–2148.

A technical paper demonstrating that volatile elements were already depleted in the planetesimals that accreted to form the planets.

Watson L. L., Hutcheon I. D., Epstein S., and Stolper E. M. (1994). Water on Mars: clues from deuterium/hydrogen and water contents of hydrous phases in SNC meteorites. Science **265**, 86–90.

An important study of water and water-bearing minerals in Martian meteorites, with implications for cycling of water through the hydrosphere.

EVIDENCE FOR LIFE IN METEORITES

Burke J. G. (1986). *Cosmic Debris: Meteorites in History*, University of California, Berkeley, CA.

Chapter 5 of this carefully researched book describes the checkered history of claims for life in meteorites.

McKay D. S., Gibson E. K., Thomas-Keprta K. L., Vali H., Romanek C. S., Clemett S. J., Chillier X. D. F., Maechling C. R., and Zare R. N. (1996). Search for past life on Mars: possible relic biogenic activity in Martian meteorite ALH84001. Science **273**, 924–930.

An intriguing but controversial paper describing putative biominerals, microfossils, and organic materials in a Martian meteorite.

McSween H. Y. Jr. (1997). Evidence for life in a martian meteorite? GSA Today **7**(7), 1–7.

This paper takes a critical view of the evidence presented for early Martian life in the ALH84001 meteorite.

Appendix of Minerals

Approximately 300 minerals have been identified in meteorites, as summarized in the references at the end of this appendix. The minerals listed below are either particularly important constitutents of meteorites or were specifically referred to in this book.

amphibole a chemically complex class of hydrous silicate minerals with similar crystal structures, orthorhombic or monoclinic, common in terrestrial rocks but only known in SNC meteorites (see **kaersutite**).

anorthite (see **plagioclase**).

augite (see **pyroxene**).

carbonates white to colored grains, solid solutions of calcite ($CaCO_3$), magnesite ($MgCO_3$), siderite ($FeCO_3$), and rhodochrosite ($MnCO_3$), intermediate compositions include dolomite and ankerite, trigonal, occur as products of aqueous alteration in carbonaceous chondrites and SNC meteorites.

chromite black oxide, $FeCr_2O_4$, cubic, minor constituent of irons, stony–irons, and some stony meteorites.

clay minerals (see **phyllosilicates**).

cohenite opaque carbide, bronze-colored, $(Fe, Ni, Co)_3C$, orthorhombic, minor constituent of iron meteorites.

corundum colorless oxide, Al_3O_3, hexagonal, occurs in refractory inclusions in carbonaceous chondrites; colored varieties such as ruby and sapphire are not found in meteorites.

diamond transparent carbon, C, dense-packed structure that forms at extreme pressures, a polymorph of graphite, produced by shock in some iron meteorites and ureilites, also found as interstellar grains in chondrites that probably formed by vapor condensation around red giant stars.

diopside (see **pyroxene**).

enstatite (see **pyroxene**).

fassaite (see **pyroxene**).

feldspar abundant class of silicate minerals, monoclinic or triclinic, includes **plagioclase** ($CaAl_2Si_2O_8$–$NaAlSi_3O_8$) and alkali feldspar ((K, Na)

$AlSi_3O_8$), occurs mainly in achondrites, as a product of metamorphism of chondrule glass in chondrites, and in silicate inclusions in irons and stony–irons.

forsterite   (see **olivine**).

garnet   complex silicate solid solution with general formula $(Ca, Mg, Fe^{2+}, Mn)_3(Al, Fe^{3+})_2Si_3O_{12}$, cubic, common in terrestrial rocks formed at high pressures, thought to be present in the source regions of SNC meteorites.

graphite   opaque carbon, C, hexagonal, sheetlike structure, stable at low pressures, occurs in some ureilites, ordinary chondrites, and iron meteorites.

ilmenite   black oxide, $FeTiO_3$, orthorhombic, common accessory mineral in achondrites and lunar rocks.

kaersutite   an amphibole, approximate composition $Ca_2Na(Mg, Fe^{2+})_4 TiAl_2Si_6O_{22}(OH)_2$, monoclinic, occurs as a minor constituent of melt inclusions in SNC meteorites.

kamacite   (see **metal**).

magnetite   black oxide, Fe_3O_4, cubic, common accessory mineral in carbonaceous chondrites and basaltic shergottites, suggested to be a biogenic mineral in ALH84001.

maskelynite   plagioclase that has been transformed into glass (substance without a crystal structure) by shock metamorphism, occurs in shergottites and some shock lunar rocks.

melilite   silicate solid solution between $Ca_2Al_2SiO_7$ and $Ca_2MgSi_2O_7$, tetragonal, occurs in refractory inclusions in carbonaceous chondrites.

metal (iron–nickel)   metallic alloys, kamacite has low (up to 7.5 wt.%) nickel, taenite has high (greater than 20 wt.%) nickel, cubic, major constituents of irons and stony–irons, minor constituents of many chondrites and some achondrites.

niningerite   opaque sulfide, (Mg, Fe)S, minor constituent of enstatite chondrites and aubrites.

oldhamite   opaque sulfide, CaS, minor constituent of enstatite chondrites and aubrites.

olivine   silicate solid solution between forsterite (Mg_2SiO_4) and fayalite (Fe_2SiO_4), magnesium-rich olivine is very abundant in chondrites, pallasites, and some achondrites and primitive achondrites.

osbornite   nitride, TiN, a rare mineral in enstatite chondrites.

perovskite   white oxide, $CaTiO_3$, cubic, a minor constituent of refractory inclusions in carbonaceous chondrites.

phosphates   white grains, commonly whitlockite ($Ca_9MgH(PO_4)_7$) or apatite ($Ca_5(PO_4)_3(F, OH, Cl)$), common accessory minerals in stony meteorites.

phyllosilicate a broad class of hydrous silicate minerals with crystal structures in the form of stacked sheets, includes submicroscopic clay minerals with complex compositions and serpentine ($Mg_6Si_4O_{10}(OH)_8$), these occur as products of aqueous alteration in carbonaceous chondrites and a few ordinary chondrites and SNC meteorites.

pigeonite (see **pyroxene**).

plagioclase an important class of silicate minerals (see feldspars) ranging in composition between anorthite ($CaAl_2Si_2O_8$) and albite ($NaAlSi_3O_8$), triclinic, a major constitutent of many achondrites and lunar rocks, as well as silicate inclusions in irons and stony–irons.

pyrite brassy sulfide, FeS_2, common constitutent of terrestrial rocks and rare phase in some SNC meteorites and a few other achondrites.

pyroxene a major group of silicate minerals with similar crystal structures, complex solid solutions between enstatite ($Mg_2Si_2O_6$), diopside ($CaMgSi_2O_6$), ferrosilite ($Fe_2Si_2O_6$), and hedenburgite ($CaFeSi_2O_6$), intermediate compositions include pigeonite, augite, and other pyroxenes like fassaite (aluminum, titanium-bearing diopside), orthorhombic or monoclinic, an important constitutent of all chondrites,

schreibersite gray phosphide, $(Fe, Ni)_3P$, tetragonal, a minor mineral in iron meteorites.

serpentine (see **phyllosilicate**).

silicon carbide SiC, does not normally occur in terrestrial rocks but industrial SiC is known as carborundum, occurs as interstellar grains in chondrites.

spinel a class of oxides with general formula $(Mg, Fe^{2+})(Al, Fe^{3+}, Cr)_2O_4$, cubic, occurs as $MgAl_2O_4$ in refractory inclusions in chondrites; **chromite** and **magnetite** are also spinels.

sulfates colorless, common varieties are anhydrite ($CaSO_4$), gypsum ($CaSO_4 \cdot 2H_2O$), and epsomite ($MgSO_4 \cdot 7H_2O$), orthorhombic or monoclinic, occur as vein materials in carbonaceous chondrites.

taenite (see **metal**).

troilite brass-colored sulfide, FeS, hexagonal, a common constitutent of most classes of meteorites.

For more information on the mineralogy of meteorites, please see the following:

Rubin A. E. (1997). Mineralogy of meteorite groups. Meteoritics Planet. Sci. **32**, 231–247.

Rubin A. E. (1997). Mineralogy of meteorite groups: An update. Meteoritics Planet. Sci. **32**, 733–734.

Glossary

absorption band a valley in a reflectance spectrum produced by absorption of certain wavelengths of energy by asteroid surface materials.

acapulcoite a type of primitive achondrite, a nearly chondritic residue from which only a small amount of melt has been extracted; related to **lodranites**.

accretion the process by which particles come together to form a larger mass of material.

achondrite a class of stony meteorites that crystallized from magmas; the term literally means without chondrules, emphasing the distinction between these meteorites and chondrites.

agglutinate a glass-bonded aggregate of broken mineral and rock fragments; a common constituent of reglith materials formed by small impacts.

albedo the percentage of incoming radiation that is reflected by a surface.

alkali anorthosite a type of igneous rock, consisting almost entirely of plagioclase, found in the lunar highlands crust.

Amor asteroid a near-Earth asteroid having a perihelion distance between 1.017 and 1.3 AU.

angrite a type of achondrite consisting of pyroxene (fassaite), olivine, and plagioclase.

anomalous iron an iron meteorite that cannot be readily classified by either its structure or chemical composition (see **structural classification, chemical groups**).

anorthosite an igneous rock consisting primarily of plagioclase; a major constituent of the lunar highlands.

aphelion the orbital position at which the distance between an object and the Sun is the greatest.

Apollo asteroid a near-Earth asteroid having a perihelion distance greater than 1.017 AU and a semimajor axis greater than 1.0 AU.

aqueous alteration transformation to a new assemblage of minerals caused by reactions with water at low temperature; this process affected most carbonaceous chondrites.

asteroid   a moving object of stellar appearance, without any trace of cometary activity.

asteroid belt   the region between Mars and Jupiter where most asteroids are found.

astrobleme   an ancient, deeply eroded crater; literally star wound.

astronomical unit (AU)   the distance from the Earth to the Sun, approximately 150 million km; commonly used to express astronomical distances.

ataxite   an iron meteorite with high nickel content, composed almost entirely of taenite and having no obvious structure.

Aten asteroid   a near-Earth asteroid having an apheion distance of greater than 0.983 AU and a semimajor axis of less than 1.0 AU.

aubrite   a highly reduced achondrite type, composed mostly of enstatite.

basalt   a common volcanic igneous rock consisting mostly of pyroxene and plagioclase.

basaltic shergottite   a class of achondrites of basaltic composition, consisting mostly of pigeonite, augite, and plagioclase; thought to be Martian rocks.

biogenic   minerals produced by organisms.

brachinite   a class of primitive achondrites consisting mostly of olivine.

breccia   a rock composed of broken rock fragments (clasts) cemented together by finer-grained material; a common product of impact processes.

carbonaceous chondrite   a clan of stony meteorites, the chemical composition of which closely matches the Sun; these are the mostly highy oxidized and volatile-rich chondrites and have commonly suffered aqueous alteration.

Centaurs   asteroids orbiting between Saturn and Neptune.

chalcophile   elements with a geochemical affinity for sulfide phases.

chassignite   an achondrite (Chassigny is the only known example) consisting mostly of olivine; related to **shergottites** and **nakhlites**.

chemical group (irons)   a classification of iron meteorites based on abundances of nickel, gallium, and germanium; groups are identified by Roman numerals and letters, for example, IAB, IVA.

chondrite   an abundant class of stony meteorites characterized by chemical compositions similar to that of the Sun and by the presence of chondrules (except CI1 chondrites).

chondrule   millimeter-sized spherule of rapidly cooled silicate melt, found in abundance in chondritic meteorites.

clan   a related group of chondritic meteorites; important clans are the **ordinary, carbonaceous,** and **enstatite chondrites.**

coma the bright cloud of gas and tiny particles surrounded a comet nucleus.

comet a body composed of ices and dust that orbits the Sun in a highly elliptical or parabolic orbit; comets may be short period (periodic) or long period.

condensation sequence formation of solids from nebular gas in response to decreasing temperature.

contact binary two asteroids orbiting so closely about each other that they may actually touch.

cooling-rate speedometer a means of estimating the rate at which a meteorite cooled, based on diffusion of nickel in metal or retention of fission tracks or radiogenic gases.

cosmic abundance the abundance of elements in the Sun; this is equivalent to the average solar system composition and is determined from chondritic abundances of all but the most volatile elements.

cosmic irradiation exposure to cosmic rays.

cosmic velocity the velocity of a meteoroid orbiting in space; this may be reduced by air friction during atmospheric passage.

crater the depression produced by an impacting projectile.

crystallization the process of producing minerals with ordered atomic structures.

cumulate an igneous rock produced by accumulation of crystals separated from a magma through some physical process.

dendrite an elongated crystal that forms from liquid.

differentiation the process by which an initially homogenous planetary body becomes internally stratified into regions of different composition; this usually produces a core, mantle, and crust.

diffusion permeation of any region by gradual movement and scattering of atoms though the substance.

diogenite an achondrite composed primarily of cumulus orthopyroxene; related to **eucrites** and **howardites**.

eccentricity the amount by which an elliptical orbit deviates from circularity.

ejecta material ejected from a crater during impact.

electron microprobe an instrument that analyzes the chemistry of very small spots by bombarding the sample with electrons and measuring the X-rays produced.

enstatite chondrite a clan of stony meteorites containing abundant enstatite and metal; these are the most highly reduced chondrites.

equilibrium   a state in which a process has produced its total effect or finished its reaction, and therefore no further change occurs.

escape velocity   the velocity that any object must achieve to escape the gravitational field of its parent body.

eucrite   a common class of achondrites, composed of pigeonite and plagioclase, that formed as basaltic flows on their parent body, probably asteroid 4 Vesta.

eutectic   the liquid that occurs at the lowest temperature in a chemical system.

exposure age   the length of time a small meteoroid (meter sized or less) was exposed to cosmic rays while orbiting in space; this is measured from the amounts of certain isotopes produced by cosmic irradiation (see **spallation**).

exsolution   the rearrangement of atoms in a homogeneous crystal on cooling, so that an intergrowth of two separate minerals results.

fall   a recovered meteorite that was observed to fall.

ferroan anorthosite   a type of igneous rock, composed mostly of plagioclase, that is common in the lunar highlands.

find   a recovered meteorite that was not observed to fall.

fireball   (see **meteor**).

fission track   a submicroscopic trail in a crystal traveled by a particle produced by radioactive decay; these can be used as a means of age determination.

formation interval   the length of time between the origin of the solar system and the formation of a particular meteorite.

fractional crystallization   crystallization in which solids are physically removed from contact with the liquid from which they grew; a common process affecting magmas.

fractionation   the separation of chemical components or minerals; sometimes used synonymously with **fractional crystallization**.

fusion crust   the glassy exterior of a meteorite; a melted rind that forms during atmospheric passage.

gas-retention age   the time at which a meteorite began retaining radiogenic gases during its cooling history; commonly used to date shock events.

GEMS   acronym for glass with embedded metal and sulfide; tiny inclusions in IDPs that may be interstellar grains.

glass   solid material without any crystal structure, produced by rapid cooling of melt or by destruction of the structure due to irradiation or shock.

group (chondrite)   a chemical classification of chondrites; groups are identified with capital letters, for example, H or CV.

half-life   the time interval required for half of the remaining atoms of a radioactive isotope to decay; this must be known for radiometric age determinations.

hexahedrite   an iron meteorite with low nickel content, consisting almost exclusively of kamacite.

highlands   the ancient crust of the Moon, consisting mostly of anorthosite and related rocks; these regions are distinguished by their high topography and heavy cratering.

Hirayama family   a group of asteroids with similar orbital characteristics; these presumably represent pieces of an earlier large asteroid that was disrupted by impact.

howardite   an achondrite breccia containing rock fragments of **eucrites** and **diogenites**.

hypervelocity impact   impact of a meteoroid traveling at greater than free-fall velocity; this produces a large crater.

inclination   the angle between an asteroid's orbit and the ecliptic plane (the plane in which the Earth orbits).

incompatible element   an ion whose size or charge is too large to fit into crystallographic sites in common minerals.

interplanetary dust particle (IDP)   a tiny particle that once orbited in interplanetary space; debris from either asteroids or comet dust.

interstellar grain   a tiny particle derived from outside the solar system.

ion   an atom with an electrical charge, produced by loss or gain of electrons.

iron meteorite   a meteorite composed primarily of iron–nickel metal (see **ataxite, hexahedrite, octahedrite**).

isochron   a straight line on an isotope evolution diagram, defined by the isotopic compositions of related rocks or mineral grains from the same sample; the slope of this line gives the age of the rock.

isotope   one of two or more atoms having the same atomic number but different mass numbers; these can be stable or unstable (see **radionuclide**).

Kakangari chondrite   a small grouplet of chondrites, with oxidation state intermediate between enstatite and ordinary chondrites.

kerogen   a complex mixture of heavy-molecular-weight organic molecules; this material is a constituent of carbonaceous chondrites.

Kirkwood gaps   voids in the asteroid belt where the orbital periods of asteroids are certain fractions of the period of Jupiter.

Kuiper belt   a region outside the orbit of Pluto thought to contain many cometary objects.

lherzolitic shergottite   meteorites composed mostly of olivine and pyroxene and related to basaltic shergottites; these are thought to be Martian samples.

light curve   changes in the measured albedo of a rotating body, due mostly to its irregular shape.

lithophile   elements with a geochemical affinity for silicate or oxide phases.

lodranite   primitive achondrite representing a residue from which modest amounts of melt were extracted; related to **acapulcoites**.

magma   molten rock materials, including suspended crystals and dissolved gases, that crystallize to form igneous rocks.

magma ocean   a huge quantity of magma thought to have existed on the early Moon; fractional crystallization of this magma produced the lunar highlands' crust and mantle.

magnesian suite   a group of old plutonic rocks that intruded the lunar anothositic crust.

main sequence   relatively young stars whose luminosity is powered by hydrogen fusion.

mare basalt   volcanic rock of basaltic composition that occurs in lunar maria.

maria   dark, generally flat areas of the Moon covered by mare basalt.

mass spectrometer   a sensitive instrument that measures the isotopic composition of a sample by separating ions of different mass using a large magnet to alter their trajectories.

matrix   dark, fine-grained material occuring between chondrules, refractory inclusions, and metal grains in chondrites.

mean-motion resonance   an orbit whose period corresponds to a simple fraction of Jupiter's orbital period; repeated close encounters of asteroids in such positions result in depopulation of such resonances.

mesosiderite   a class of stony–iron meteorites consisting of metal and fragments of igneous rocks similar to eucrites and diogenites; these formed as breccias, but most have been recrystallized during metamorphism.

metal   metallic iron–nickel alloys commonly found in meteorites.

metallographic cooling rate   the rate at which a meteorite cooled through a temperature of approximately 500 °C, calculated from measured diffusion profiles of nickel in metal.

metamorphism   recrystallization, in the solid state, of a rock in response to high temperature or sometimes pressure.

meteor   a streak of light in the sky produced by transit of a meteoroid through the Earth's atmosphere; also, the glowing meteoroid itself.

meteor stream   a group of meteoroids following the same orbital path and presumably once part of the same object.

meteorite   extraterrestrial object that survives passage through the atmosphere and reaches the Earth's surface as a recoverable object.

meteorite stranding surface   a surface on an icy substrate where meteorites are concentrated.

Meteoritical Society   an international organization of scientists and meteoritophiles devoted to the study of meteorites, other extraterrestrial materials, and planetary science; this organization approves names of new meteorites.

meteoroid   a small object orbiting the Sun in the vicinity of the Earth.

micrometeorites   small particles, derived either from asteroids or comets, that generally melt or are vaporized during atmospheric passage.

nakhlite   a class of achondrites composed primarily of augite and olivine; related to **shergottites** and **chassignites** and thought to be Martian samples.

near-Earth object (NEO)   an asteroid or comet orbiting near the Earth (see **Apollo, Aten**, and **Amor asteroids**).

Neumann lines   twin boundaries in metal grains of iron meteorites formed by shock deformation.

nucleosynthesis   the formation of elements, by a variety of nuclear fusion processes in the interiors of stars or supernovae.

nuclide   (see **isotope**).

octahedrite   an iron meteorite of intermediate nickel content, containing both kamacite and taenite in a Widmanstaettian pattern.

onion shell model   a proposed thermal model for chondrite parent bodies in which various petrologic types (metamorphic grades) are arranged concentrically within the body.

Oort cloud   a massive swarm of comets thought to surround the solar system in a spherical volume extending out to at least 50,000 AU.

orbit   the elliptical path followed by one object revolving around another.

ordinary chondrite   the most common class of stony meteorites; this clan consists of several chemical groups – H, L, and LL.

organic matter   compounds of carbon, hydrogen, and commonly oxygen, nitrogen, sulfur, and phosphorus; these typically form complex molecules.

oriented meteorite   a meteorite of conical shape, sculpted during transit through the Earth's atmosphere.

oxidation   loss of electrons resulting in the conversion of reduced species like iron metal into oxidized species like ferrous and ferric ions.

paired meteorites   two or more separate meteorites that are parts of the same fall.

pallasite   a class of stony–iron meteorites consisting of metal and isolated crystals of olivine.

panspermia   the hypothesis that life originated elsewhere and the Earth was innoculated with extraterrestrial organisms.

parent body   an object of asteroidal size or larger from which meteorites were derived.

partial melting   the mechanism by which preexisting rocks give rise to magmas; complete melting requires temperatures that are usually so high that it is rare in nature.

period   the time required for an orbiting body to make one complete revolution.

perturbation   any disturbance or minor sinuosity imposed on an otherwise elliptical orbit; this is usually caused by gravitational attraction of a third body.

petrographic microscope   a microscope used to examine thin sections of rock; the interaction of minerals with polarized light serves to identify them.

petrologic type   a relative scale from 1 to 6, reflecting the degree of thermal metamorphism or aqueous alteration in chondritic meteorites; relatively unaltered meteorites are type 3, with metamorphosed samples at high numbers and altered samples at lower numbers.

phase diagram   a diagram showing the stability fields of various minerals in terms of temperature, pressure, or composition.

planetesimal   a small orbiting body; usually synonymous with asteroid.

plutonic   igneous rocks that crystallized slowly within the interiors of planets or asteroids.

polycyclic aromatic hydrocarbons (PAHs)   complex organic molecules consisting of fused rings of carbon.

primitive achondrite   a meteorite formed as a residue from which small amounts of melt were extracted; the major element composition of these meteorites remains chondritic, but their textures and trace-element abundances are usually modified by heating and melt loss.

prograde orbit   a path in which the orbiting meteoroid revolves around the Sun in the same sense as do the planets.

pyrolysis   reactions among organic matter that result in loss of water.

racemic mixture an equal mixture of left- and right-handed stereoisomers; abiotic synthesis normally produces racemic mixtures.

radar an intense, coherent radio signal; by comparing the echo's characteristics with those of the transmitted waveform, the sizes, shapes, and surface properties of asteroids can be determined.

radioactive unstable nuclide that undergoes spontaneous decay to produce an isotope of another element.

radiogenic an isotope produced through radioactive decay.

radiometry the study of the thermal brightness of asteroids, measured from infrared radiation.

radionuclide an unstable (**radioactive**) isotope.

rare-earth elements lanthanide series of the Periodic Table; closely related elements with large ionic radii that are useful in geochemical modeling.

reflectance spectra (see **spectrophotometry**).

refractory inclusion white inclusion, commonly irregular in shape, that occurs in chondrites; these are thought to be condensates or residues from partial evaporation that formed in the solar nebula; also called calcium–aluminum inclusions or CAIs.

regmaglypt depression resembling a thumbprint that is produced on the surfaces of some meteorites during atmospheric transit.

regolith a layer of fragmental, incoherent rocky debris that nearly everywhere forms the surface terrain; produced by repeated meteorite impacts.

regolith breccia compacted and lithified regolith material.

residue the solid material remaining after devolatilization or extraction of a partial melt.

retrograde orbit an orbit in which the body revolves around the Sun in the direction opposite to that of the planets.

r process a rapid fusion process leading to nucleosynthesis of heavy elements; this process occurs during supernova events.

rubble-pile model a model for chondrite parent bodies in which catastrophically disrupted asteroids reaccrete; thus petrologic types are randomly distributed within the body.

Rumuruti chondrite a small clan of chondritic meteorites, resembling ordinary chondrites but more highly oxidized.

s process a relatively slow nucleosynthetic process involving the production of new elements by nuclear fusion in red giant stars.

secular resonance a situation in which the rate of precession of the node or perihelion of an asteroid's orbit is equal to that of a major planet; over long periods of time, asteroids near a secular resonance experience large perturbations.

semimajor axis   the long axis of an ellipse; one characteristic of orbits.

shergottite   (see **basaltic shergottite, lherzolitic shergottite**).

shock metamorphism   the effects of high pressure due to impact into a target rock; these effects include transformation of mineral structures, brecciation, melting, and other modifications.

siderophile   elements with a geochemical affinity for metallic iron.

silicate inclusions   pieces of stony meteorite contained within iron meteorites.

snow line   the distance from the Sun, approximately 2.5 AU, at which volatile ices first condensed in the solar nebula.

solar nebula   the disk-shaped cloud of gas and dust from which all bodies in the solar system originated.

space weathering   processes thought to modify the spectral properties of materials on the uppermost surface layer of an airless body.

spallation   the production of new isotopes through irradiation by cosmic rays.

spectrophotometry   the study of the spectrum of sunlight reflected from the surface of an asteroidal or planetary body; spectra can be related to mineralogy.

stellar occultation   the cutoff of light from a star due to its passage behind an asteroid; a method used to determine the sizes of asteroids.

stereoisomers   organic molecules having the same chemical composition but mirror-image geometries; these are normally referred to as left and right handed, based on which way they rotate polarized light.

stony–iron meteorite   a meteorite consisting of approximately equal parts silicate minerals and iron–nickel metal (see **mesosiderite, pallasite**).

strewn field   an elliptical pattern of distribution of recovered meteorites, formed when a meteoroid is fragmented during atmospheric passage.

structural classification (irons)   a classification of iron meteorites by their observable structural features; recognized groups are **ataxites, hexahedrites**, and **octahedrites**; the latter are further distinguished by the widths of kamacite bands.

sulfide   a mineral formed by combination of a cation with sulfur.

supernova   the explosion of a star; this results in the production of new elements and their dispersal into interstellar space.

terminal velocity   the free-fall velocity of a meteoroid due to the Earth's gravitational attraction after its cosmic velocity has been slowed by atmospheric friction.

terrestrial age   the length of time a meteorite has resided on Earth, measured from the decay of radionuclides produced by cosmic-ray exposure in space.

thermodynamic calculations   theoretical calculations based on relations between heat and mechanical energy; a means of estimating the condensation sequence in the solar nebula.

thermoluminescence (TL)   the thermal release of energy stored in crystals by ionizing radiation in space; a measure of cosmic irradiation.

ultramafic   rocks composed primarily of minerals rich in magnesium and iron, such as olivine and pyroxene.

ureilite   an unusual class of achondrites consisting mostly of pigeonite set in a carbonaceous matrix; diamonds occur in shock-metamophosed ureilites.

volatile element   an easily volatilized substance that condenses from a gas at low temperature.

volcanic   an igneous rock that formed by rapid solidification of magma that erupted onto the surface of a planet or asteroid.

Widmanstätten pattern   a regular geometric intergrowth of kamacite plates within taenite that occurs in some iron meteorites (see **octahedrite**).

winonaite   a class of primitive achondrites closely related to some groups of iron meteorites.

xenolith   an inclusion of foreign rock trapped within an igneous rock.

Index